GENDER AND TECHNOLOGY
IN THE MAKING

GENDER AND TECHNOLOGY IN THE MAKING

Cynthia Cockburn
and
Susan Ormrod

Photographs: Cynthia Cockburn

SAGE Publications
London • Thousand Oaks • New Delhi

First published 1993

 SAGE Publications Ltd
6 Bonhill Street
London EC2A 4PU

SAGE Publications Inc
2455 Teller Road
Thousand Oaks, California 91320

SAGE Publications India Pvt Ltd
32, M-Block Market
Greater Kailash – I
New Delhi 110 048

British Library Cataloguing in Publication Data

Cockburn, Cynthia
  Gender and Technology in the Making
  I. Title
  604.8

  ISBN 0–8039–8810–9
  ISBN 0–8039–8811–7 (pbk)

Library of Congress catalog card number 93–085147

Typeset by Mayhew Typesetting, Rhayader, Powys
Printed in Great Britain by The Cromwell Press Ltd,
Broughton Gifford, Melksham, Wiltshire

TO THE MEMORY OF SALLY HACKER
Thanks for the fun, the insight and the anger

Contents

Acknowledgements

Many people contributed to this book. The research on which it is based we carried out together over two years from mid-1989 to mid-1991. Cynthia wrote the original research proposal, had overall responsibility for the project and did some of the interviewing; Susan carried out most of the interviews, wrote interim reports and contributed substantially to the analysis and theorizing of the findings. She then moved on to other work and Cynthia remained to write the book, helped by her comments and suggestions.

We would like to thank the companies that afforded us research access and all those in them who gave their time to help us in interview or by providing information in other ways. Unfortunately the requirements of anonymity do not permit us to name you. This applies also to the women, men and children interviewed or photographed at home, whom we thank warmly for their patience with our questions and cameras.

We are very appreciative of the financial support afforded by the Economic and Social Research Council (R 000 23 2249), which funded our project in full, and the administrative support of The City University, where the project has been located in the Centre for Research in Gender, Ethnicity and Social Change.

Thanks are due too to the European Centre for the Co-ordination of Research and Documentation in the Social Sciences, and most particularly to Ruza First-Dilic, for coordination of the international group in the context of which this research was carried out. The many discussions held with members of this nine-country group over a period of four years have helped develop both the research method and the analysis.

Especial thanks are due to the following who read and gave valuable comments on the draft, in part or in full: Anne-Jorunn Berg, Sandra Harding, Trevor Pinch, David Rea and Judy Wajcman. Louise Murray and Karen Phillips at Sage have been encouraging and helpful editors. If the book finally has anything to commend it, it is due to the help we have received from others. Its weaknesses are not of their making. Finally Cynthia Cockburn wishes to thank the Disablement Advisory Service for lending a Dragon Dictate speech-recognition word processing system on which this book was 'typed' in hands-free mode, and Aptech Ltd for their technical support in its use.

Introduction

An effective introduction to the curious relationship of gender and technology is to reconsider that old chestnut 'the woman driver'. Out there in the busy street, a stream of traffic. Somebody does something inept, stalls the motor on a hill, takes two attempts to park. Other drivers smugly look to see who is the incompetent. At the wheel they see (as it happens) a female. Ah! 'Woman driver!' Even women find themselves saying this, though the feelings with which they discover the fact and speak the words may well be more ambivalent than those of a man.

The theme of this book could be summed up as exploring the several truths about gender and technology nested within this kind of phenomenon. First, in there somewhere are material, even quantifiable, systematic differences between the sexes in our culture. Women are practically distanced from certain technologies in their upbringing and the work they typically do. On average they acquire less early experience of certain machines (let us say driving and owning their own car) than do men. Women drive less often, for shorter distances. This material difference is relational: women driving less has to do with men driving more. As a result of their history women are often, in truth, less technically competent in control and maintenance of, for instance, vehicles as a result.

There are also, however, issues of representation and meaning here. There is a prevalence of bald prejudice: women are expected and believed to understand and control engineered technologies less competently than men. We are *ready* to blame a 'woman driver'. Further, our differences of experience, the words we hear and the images we see help produce our self-identity. The 'woman driver' cliché, so often ringing in our ears, helps to produce an unconfident woman driver and contributes to the making of men in the persona of 'driver', a driving person.

Finally, the meanings we exchange among ourselves generate a hierarchy. Certain technologies are given higher status than others. Driving a powerful automobile and driving it 'well' are considered important personal attributes. It is laughable to think of control of a washing machine in such a light. The hierarchy so created is a gender hierarchy. It is not incidental that the former technology is associated with men, the latter with women. If in some sense we 'gender' artifacts, so also do we gender skills. Women as a sex drive 'better' than men in a certain sense of the word, less aggressively and less fast. Yet such performance is not counted as skill: *men's* abilities are skill. Yet (the dialectical double 'yet') women becoming practised drivers and motor mechanics – and indeed critics of the internal combustion engine – can begin to shift meanings and values.

In the 'woman driver' effect we are seeing something of *the gendered relations of technology*. This book is not about automobiles. Insofar as it is about one artifact it is about microwave ovens. But one technology would do as well as another, for the book is centrally about the *technology/gender relation*. It explores some technological processes in order to learn more about the disadvantage of women, the relation of women to men, of the feminine to the masculine. It reveals the disadvantage and subordination of women in which these sets of relations are implicated, and opens up a prospect of resistance and change. In doing these things it also tells us something new about technology.

Since the revival of the women's movement in the late 1960s, the relationship of women with technology has been consciously problematized. Feminist research on occupational sex-segregation has revealed the significance of technology and technical knowledge and know-how: jobs calling for these qualities are mainly done by men, and seen as masculine jobs (Bradley, 1989; Game and Pringle, 1983; Walby, 1988 – these and following references are just a few examples drawn from many studies). Educational sociologists have demonstrated the way girls and boys opt for their school subjects, perceive the world of work and often make sex-typed occupational and training choices (Chisholm and Holland, 1986; Cockburn, 1987; Deem, 1986; Holland, 1986, 1987; Kelly, 1981). Psychologists have probed the male and female psyche for indications of gender difference in regard to engineering (Newton, 1987; Weinreich-Haste and Newton, 1983) and sociologists have uncovered the contradictory experiences of the few women who do become engineers (Carter and Kirkup, 1990). The research, which set out with a sense that women and girls might be 'failing' to grasp the significance and potential of engineering skills, moved quickly to a perception that they often positively chose something else, something more human. Likewise the studies of sexual divisions in employment moved on from detecting employers' and trade unions' frankly exclusionary practices against women to tracing the more subtle cultural processes in which we could see the so-called skilled sphere being appropriated for masculinity, while femininity was constituted around a lack of recognized skill (Cockburn, 1983, 1985; Hacker, 1989; Phillips and Taylor, 1980). A systematic undervaluation of women's particular rationality and expertise was laid bare (McNeil, 1987).

Feminist research on technology has been usefully reviewed and assessed by Judy Wajcman (1991), who confirms that by the late 1980s attention in feminist technology studies was tending to shift away from the focus on 'women and technology'. Instead it was beginning to examine the very processes in which technology was developed and used, and those in which gender was constituted. In that respect this research project and this book are children of their time. It was in 1989 that the European Centre for Co-ordination of Research and Documentation in the Social Sciences (the Vienna Centre) invited women from ten countries of Europe to carry out cross-national research on 'the impact of changing technology on gender relations'. It is symptomatic of the moment that some of us in the group felt happier reformulating this as: 'the mutual shaping of technology and gender relations'. There was a feeling from the time of our first discussion, that we would need to be asking not only 'How does technology bear on gender relations?' but also 'How are technological outcomes shaped by gender?'

We decided to focus each of our various country research projects on a different artifact, each one destined to end up in the home, in domestic use. This was a tactical device for ensuring that the study of technology would project us into an encounter between masculine and feminine spheres. We also decided to focus our studies not on any one moment in the life of a technology (design, diffusion etc.) but rather to trace the whole life trajectory of an artifact. The reason for this methodological choice was theoretical and will become apparent. The report on the collective project will be published as C. Cockburn and R. First-Dilic (eds), *Bringing Technology Home: Women, Gender and Technology in Europe*, Open University Press, 1994.

In the British project we selected for study the microwave oven. As a technology, it seemed recent enough to count as an innovation, yet established enough to furnish some popular experience to study. It would enable us to observe an industrially engineered technology, with a new heat-generating principle, impinging on an older 'technology': domestic cooking. Stretching back to the wood fire and the clay pot, this one had traditionally been in women's hands.

Focusing our research on an artifact, the hard, tangible, rectilinear microwave oven, may seem to reify technology unduly. The choice can however be seen differently. The artifact was no more than a device to draw us into many different places and involve us in many different activities where the technology/gender relation is enacted. The artifact provided a rationale for, and gave coherence to, a sequence of contacts and case studies.

As is common in qualitative research, our sources of information were several. We drew on current theory. We sought documentation wherever we could find it, from trade reports and company publicity to household magazines. We 'hung out' in research sites, observing environments and interactions. We engaged in phone interviews, and in informal discussions. A research collaborator, who happened also to be an experienced electrical goods salesperson, provided us with a useful insider view from two months' participant observation in a sales job in a retail store. We carried out a mini-survey by questionnaire of microwave owners, with 34 usable replies.

Our most important source of information, however, was interviews with 89 people. These interviews were semi-structured, tape-recorded and involved in total 43 women and 46 men. Twenty-seven of them were involved in the case study of the firm manufacturing microwave ovens which we call Electro. A further 21 interviews were carried out in two retail chains. We draw in the book mainly on one of these (calling it Home-Tec), though the information obtained in the second of the case studies (Wonderworld) is drawn on where it amplified our understanding of the process in which microwaves are retailed. We have purposely suppressed or reshaped key facts about our case-study firms, at their own request, to assure their anonymity. Twenty-five of the interviews were carried out in 18 households using microwave ovens. Of these, five were Midland family pubs – but that material is not discussed here. The remaining interviews involved individuals and organizations implicated in the life trajectory of the microwave in other ways: home economics teachers, advertisers, service agents.

The use of photography calls for some comment. Cynthia Cockburn decided to return to research sites with a camera when she was first impeded in writing by 'repetitive strain injury' from too many years of typing. No ambitious claims are made for the photographs from a sociological point of view. They clearly do not represent some definitive 'reality': they are highly subjective and selective. Nor do they set out to illustrate our findings in some straightforward way. They are offered simply as a parallel narrative, with resonance in the text. They may prompt some thoughts in the reader that words alone could not evoke about the relationship of women and men to artifacts and, in the context of technology, to each other.

It will be evident that our choice of method – linked case studies, phases in the life trajectory of the artifact, involving an overview of a wide network of actors and agencies – had its cost as well as its advantages. In a project involving only two people for two years, we were thinly spread. At no point did we feel satisfied that we had all the information, all the depth of insight, we needed – or which, given more time and more resources, we might have obtained. On the other hand, we felt throughout that the approach had a pleasing innovatory quality and was producing an understanding of the interrelationship of technology and gender that no other research design could have enabled.

In any case, no one research approach could possibly, in the very nature of things, be expected to generate a definitive account. This, like any other piece of social analysis, has to be seen as one telling of a story of technology and gender, an account contingent on both the technology chosen and the theoretical starting point, and taking a slant from being the perspective of two feminist researchers, British, white, working in an academic context. Our best hope for it is that it will be found to have some resonance with tales told by other women and men in various relationships to technology.

A Way of Seeing Gender

The everyday way of understanding the differences and inequalities between women and men is to put them down to 'Nature'. People say, 'We were made different'. They say, 'Boys will be boys!', or 'Vive la différence!' There is indeed a school of sociobiology that seeks to legitimize this popular view (Wilson, 1975). Both are contradicted by feminism, which asserts that women are not destined by their biology to certain roles and behaviours, that gender differences and inequalities are a matter not of nature but of *culture*. When sociology appropriated the term 'gender' from grammar, where it had more commonly been used, it was to enable precisely this distinction to be made between biological sex differences and the many more elaborate differences between women and men that are socially constructed (Oakley, 1972: 158). Culture responds to our biological dimorphism, now exaggerating it, now negating it, always transcending and transforming it. It is in this social sense that we use the term 'gender' in this study. We identify two sexes unproblematically – as all our research informants do – speaking of 'men' and 'women'. Yet we would wish to leave open just what those categories mean to different people in different contexts, or what they might come to mean.

A great deal of evidence supports the feminist case against biological determinism (Epstein, 1988). Science itself shows that being born with or without the Y chromosome has little or no direct effect on whether persons *feel* that they are female or male (Kessler and McKenna, 1978). Many anthropological and sociological studies have demonstrated the wide cultural variation in notions and practices of gender difference (Leacock, 1981; Mead, 1935; Ortner and Whitehead, 1981; Rosaldo and Lamphere, 1974) and have analysed the processes in which gender is actively constructed (Lorber and Farrell, 1991). One particularly influential stream of work has focused on the special significance of language in the forming of gendered identities (Weedon, 1987).

In its more visible forms, the feminism of the late 1960s and 1970s had been mainly the product of white middle-class higher-educated Western women. This hegemony was challenged increasingly during the eighties by black women, Third World women and working-class women in the industrialized countries, each claiming the validity of their own standpoint, expressing related but different oppressions, engaged in related but different movements of liberation. Feminism had exposed the narrow partiality of knowledge claims by white, ruling-class men passing for universal truth (Pateman, 1988; Lloyd, 1984). Now feminism itself acknowledged that for white Western women to claim to speak for 'women' was to marginalize and silence many other groups (Fraser and Nicholson, 1990; Haraway, 1991; hooks, 1982). Our use of 'gender' here responds to this perception. We see masculinity and femininity as plural phenomena, the forms of male dominance changing over time and differing in different cultures.

Certain things about gender are constant across cultures, however. We can say, for instance, that it is universally a relation and a process. What further general understanding of gender could we carry into this study of the social shaping of a technology? Sandra Harding has discussed in some depth the gendering of science relations (Harding, 1986). We found that the articulation of gender she developed in that context could appropriately serve as a starting point in our study of technology.

Harding distinguishes three aspects of gender. She identifies, first, *gender structure*, by which she means the sexual division of labour; second, *gender identity*, or *individual gender*; and third, *gender symbolism*, 'a fundamental category within which meaning and value are assigned to everything in the world' (Harding, 1986: 57). If we neglect one or another of these faces of gender the result is a failure of strategy for women in relation to science. The same applies, we believe, to women in relation to technology.

In the account that follows, therefore, we too use the concept of *gender structure*, and show such structure as existing both within and beyond the immediate world of the microwave oven. Women and men are situated in sex-typed ways in relation to jobs in the microwave manufacturing plant, the shops that sell microwaves, the households that use them. But these women and men walk into our microwave story already-women and already-men, bringing with them the effects of a lifetime lived in the family, the street, the school, the shopping precinct, the pub. Those gender-structured experiences in the wider society propose and dispose, without determining, their actions within the microwave-world.

Like Harding, we see gender structure as articulated with hierarchical structures of class and race, as well as having other dimensions. We also go beyond the 'sexual division of labour' to emphasize contemporary practices in which the sexes tend to be differently positioned and to behave differently, but which cannot accurately be called 'labour'. For instance, in our microwave studies, women and men play differentiated parts in such activities of family life as caring, deciding, planning, playing and resting. We have therefore devised the term *gender pattern of location* to express this structuring of work-and-other-activities as we found it, following the microwave around its life circuit.

It must be stressed that we do not see social structure as ever in any sense determining, and human beings may never be reduced to mere 'bearers' of structure. These inherited or contextual structures or patterns are continually in process. The sexual division of labour, the distribution of wealth, the rules governing citizenship, for example, do not stand still. They change as women seek new work, as governments rewrite tax or immigration laws. Structures are continually being 'remantled', reconstructed, adapted and renewed through individual and collective action.

Secondly, we use the concept of gender identity, distinguishing further between *projected identity* and *subjective identity*. By the former we mean potential, actual or desired gender identities as others perceive or portray them (for instance, the identity of 'the housewife' as constructed by the appliance designer or the advertiser). By the latter we mean the gendered sense of self, the identity created and experienced by the individual.

Thirdly, in adopting the notion of gender symbolism as the third face of gender we develop it in terms of *representations* and *meanings*, looking in particular for ways in which gender gains expression in technology relations, and technology acquires its meaning in gender relations. We find the continual *asymmetry* underlined by Harding: masculine and feminine exist always in relation to each other but are never equivalent and cannot be treated as though they were. To construct the masculine is to differentiate it from something of lesser value called the feminine.

Finally, we emphasize the mutual support of these three aspects of gender, the way they interact to sustain asymmetry and male dominance, but also the *tension* between them, the discrepancies, the contradictions that arise between them over time and open up possibilities for feminist change (Harding, 1986). Feminist postmodernism, particularly in the context of literature studies, sometimes represents culture as discourse, discounting the material. Seen in this way, gender is a discursive practice or 'performance' (Butler, 1990). A study of *technology* and *food*, however, presses on us a consideration of material as well as representational factors. We shall see many examples of a recursive relation between the material and the representational. Representations shape material practices (to be *told* that engineering is a job for men increases the percentage of engineers who are male). But the material is itself a source of meaning (if I see that of ten engineers nine are men, this *tells* me something about both engineering and men). It is precisely the interrelation between the material positioning of women and men and the meanings that are generated by and about it that is the focus of this book. We shall see how, when the locations of a few women and men shift, the masculine and feminine as symbolic spheres adapt and continue, limiting the transformative effects of material change.

A Way of Seeing Technology

Starting a study of the gender/technology relation with this kind of perception of gender has the effect of posing certain requirements of a theory of technology. The approach to technology we use here is derived from the school of thought that represents technology as being, like gender, first and foremost *social*. The sociology of technology has been an active field in the last ten or twelve years. While a mainstream body of work has emphasized that innovations are 'socially shaped', feminists have further demonstrated that this 'social' in which the shaping occurs, often interpreted as class relations, is also a matter of gender relations.

Within the thinking on 'the social construction of technology' there are differences of view as to the significance of material factors. Some assert the totality of the social: no hard, purely technical, core remains to technology when, one after another, the social skins of the onion are peeled away (Grint and Woolgar, 1992). In studying the microwave we have found it more useful to see human actors (as actor-network theorists do) as *struggling with* the material: geographical factors, radiation, bacteria (Law, 1989).

The position from which today's social shaping ideas depart is technological determinism. In this there is a parallel with sex/gender and biological determinism – less coincidental than due to overall trends in social theory. The determinist view sees technology as an autonomous force 'impinging on society from outside of society' (in the words of a useful critique by MacKenzie and Wajcman, 1985: 4). As with biological determinism, technological determinism is the popular way, even today, of understanding technology. People say casually, for instance, 'the micro-chip is causing mass unemployment', or, 'technology is out of control'. In this rendering of the history of technology its origins either lie in individual inventive genius or flow inevitably from the current state of scientific knowledge. Once the structure of the atom is understood the Bomb will ensue. Either way, whether it needs a great mind or a prior discovery, a new technology is seen as something whose 'time has come'. Viewed this way, technology seems beyond our influence. All we can do is make impact studies and dwell, often appalled, on its effects.

Determinism, whether biological or technical, is a conservative force tending to preserve existing power relations and disguise the possibility for social change. In technological studies, determinism has increasingly been countered by the idea that successive technologies are the choices of groups and forces whose interests and purposes can be identified and demonstrated: capitalist employers, the industrial/military complex, the state (Albury and Schwartz, 1982; Lazonick, 1979; Noble, 1977, 1984; Winner, 1980). Trajectories can be seen to vary according to economic, political and social context (MacKenzie and Wajcman, 1985). Feminist researchers have shown that masculine interests too are identifiable in technical choices, often overriding or subordinating the interests of women (Berg, 1990; Cowan, 1985; Cockburn, 1983). This has marked a shift from a mesmerized gaze on the ineluctable effects of technology to an investigative and sceptical concern with human actors.

In the mid-1980s a fruitful convergence occurred between such social studies of technological innovation and already well-developed theories concerning the social nature of scientific 'discovery' (Bloor, 1976; Collins, 1981). It began to be understood that the human activities in which artifacts originate do not differ greatly from those in which ideas are generated. Since scientific facts and technological artifacts are both social constructions they may be studied using similar concepts (Pinch and Bijker, 1990). Particularly influential have been those arguing the merit of detailed laboratory studies of 'science in action' using the concept of the 'actor-network' (Callon, 1986; Latour and Woolgar, 1979; Latour, 1987).

Some of the concepts developed in the social constructivist and actor-network studies have furnished us with a language to use in our study of the microwave oven. First, science and technology are *culture*. Ideas and artifacts are social constructs, the outcome of negotiation between *social actors*, both individuals and groups. To explain a technological development we need to identify the people involved, observe what they do, what they say and how they relate. A successful innovation, like microwave cooking, depends on the creation and maintenance of an *alliance* of actors. Alternative scientific findings and competing technological projects are continually being hypothesized or assayed: a project has a certain *multidirectionality*. Some fail, a few succeed. The selection of one of many possible directions is a social process. Just as scientific findings are usually open to more than one interpretation, so technological projects too can be taken up and developed, and artifacts used, in more than one way: there is a degree of *interpretive flexibility*. A technology's consequences therefore may be unintended, surprising. Its impacts cannot always be read off from the interests of its originators. Finally: *closure*. A successful mobilization of arguments or resources will, for a while, win the day. An idea will be accepted as true, an artifact will enter production. Each can however expect to be challenged, amended, remodelled – and again closed. We shall see all these things happening in the case of the microwave oven.

Though owing much to this mainstream of social studies of technology, our microwave study is obliged to diverge in several ways. This is because our main aim is different: we are seeking to understand one specific aspect of technology relations: their *gendering*.

For one thing, we have to find the women. Social studies of technology have, in the main, been concerned with the initiatory moment – invention, innovation. The principal actors therefore have been scientists and engineers, and those entrepreneurs and authorities they must draw into an effective alliance if their technological project is to move forward. Such is the sexual division of labour that few if any women are to be found among these actors. The white-coated scientists, the hard-hatted engineers, the grey-suited business executives – these are almost all men. Women are invisible in the mainstream technology studies partly because of their actual absence from the network as there defined.

There is a relatively simple corrective: extend the scope of the technology world. We had chosen an artifact already well launched into a cycle of production, marketing, consumption, feedback, redesign. We had only to shift our gaze beyond the walls of the design office and many women came into view as actors. We would pay attention to production (for many women are found among assembly-line workers manufacturing technological products), to distribution (where they are often the sales staff and the shoppers), and to the household (where they are the majority of users of domestic technologies). With certain exceptions (such as Pinch and Bijker, 1990) social shaping studies underplay the importance of enrolling groups such as manufacturing operatives and high street consumers in the alliance of forces that enables a technological innovation to succeed. There is a valid political case, beyond that of women, we believe, for seeing limitations in closely focused innovation studies. Langdon Winner has pointed out that their preoccupation with powerful and pro-active individuals and groups tends to lead to a neglect of the part played by *all* marginalized and subordinate groups (Winner, forthcoming).

It is not enough, however, simply to 'add in women'. We have to tackle the prevailing blindness to gender itself – including the neglect of the masculinity of male actors. These men too have to be seen as gendered individuals, relating in gendered ways. We have to ask different *kinds* of questions about the actors, relational questions. Because our study is concerned with gender, it inevitably has this dimension of *subjectivity*, of *identity*. It deals with feelings, aspirations, resentments: with the personal.

It also, conversely, has to connect with the more far-reaching phenomena of *gender structure*. We have, as mentioned above, found it useful to maintain a distinction between local action and individual agency on the one hand, and, on the other, the longer-lived and more widely spread social structures, particularly those of class and gender, that shape probabilities and incline or dispose our individual and collective choices of behaviour and thought. In this use of social structures or patterns of relationship we diverge from some 'social shaping' theories, in particular actor-network approaches, in which there is agnosticism about social structures existing outside or prior to the interactions observed. Rather than seeing the macro-structures of the wider world sometimes influencing events in the laboratory, they prefer to focus on the way successful micro-actors can be observed to create the alliances that enable them to grow to macro size. They argue against explanations based on imputed 'interests', on 'unexamined phenomena' or 'unexplicated resources' (Callon and Latour, 1981; Callon and Law, 1982; Woolgar, 1981; Latour, 1986). Sandra Harding has pointed out that actor-network theory's insistence on the isolation of research communities from the larger social, economic and political currents in their societies is a flaw, a kind of positivism (Harding, 1991: 162). To avoid confusion and to signal the fact that we fail to observe some of the methodological rules of actor-network approaches we have chosen not to use the, otherwise very apposite, term 'network' in our own study. Instead we talk of our 'actor-world' or 'microwave-world'.

In this microwave-world, besides, we do not hesitate to suppose the conduct of managers may sometimes be argued as expressing the 'interests' of a profit-oriented industry, and that of individual men as sometimes expressing the 'interest' of men as a sex. We would argue (with Donald MacKenzie) that the imputation of social interests to social structures and institutions, though always contestable, is both necessary and legitimate. Contemporary laboratory studies may be able to eschew external 'interests', MacKenzie pointed out, but historical explanation routinely and inevitably involves invoking causes not identifiable independently of their effects (MacKenzie, 1981).

We had a particular purpose in combining an *innovation* study of the microwave oven with a *user* study of microwave cooking practice, and indeed in going beyond this dichotomy by taking as our subject the whole life-trajectory of our artifact, including the moments of marketing, sale and purchase. We believe that both innovation studies and impact studies, alone, have shortcomings.

Impact studies have all too often been associated with technological determinism. They have examined effects without questioning causes. On the other hand the shift of focus away from impact to innovation studies has led to a forgetting of some of those important questions that used to be asked about the effects of technology on ordinary people. The result, as Langdon Winner has pointed out, has been a loss of political awareness (Winner, forthcoming).

Innovation studies leave unanswered the question: how much real interpretive flexibility exists once a given commodity (artifact, appliance) reaches the hands of the consumer? They may lead to an underestimation of the extent to which a telephone answering machine, a dishwasher, a microwave oven can be used in alternative ways, for alternative purposes and with different effects from those foreseen by their manufacturers. In the case of such domestic appliances it is certainly important to be open to this possibility, since it is one of few potential sites for women to be creative participants in technological innovation.

On the other hand, to overestimate the interpretive flexibility available to the consumer may lead us to be complacent about the direction of technological change. Rob Kling has demonstrated how a shift of attention from production to consumption can remind us that new products are all too often the subject of uncritical celebratory discourses by business interests that ought to be challenged (Kling, 1991, 1992). He points out that technologies and related social arrangements enable, facilitate, inhibit or catalyse other social changes (Kling, 1992: 359). We shall certainly see this in the case of microwave cooking. The 'interpretivism' often associated with innovation studies runs the risk of underestimating the effect of technology as a discrete independent variable in shaping work and everyday life. 'It is one thing to say that technology is not the prime determinant – it is quite another to say that it has no impact at all' (Rose et al., 1986).

The Technology/Gender Relation

It is significant that the few in-depth *gendered* case studies of technologies and technical change published in recent years have been by women and have been carried out at some conceptual distance from the (almost-all-male) sociology of science and technology referred to in the previous section. A medium commonly chosen by women for such studies has been history, through which they have reconstructed a series of instances in which technologies can be seen coming into existence and into daily use, shaped by and shaping gender relations.

Louise Walden for example has told the story of the sewing machine, the first mechanical device to be manufactured on a large scale in Swedish factories and the first industrial-technological appliance to be brought into the homes of all classes. She shows how this innovation, the brain-child of a man, constructed, produced and sold by men, became a new means of women's labour in the household (Walden, 1990). Ruth Schwartz Cowan in a history of three centuries of household technologies in the USA shows how economic forces shaped the private family house-hold and housewife, and how male inventors and designers produced one generation after another of domestic technologies for women's use (Cowan, 1989). Michele Martin traces the early years of the Bell Telephone system in Canada. She shows how the company employed women as switchboard operators, consciously using their femininity, but also how women as telephone users influenced the development of telephone systems. Women's use of the phone for sociability rather than business eventually persuaded a reluctant telephone corporation to conceive of its networks as a residential service (Martin, 1991).

In a fourth historical study, Miriam Glucksman, like Martin, intro-duced women *workers* as actors in the technological systems. The technological innovation that is the focus of her study is not a product but a process technology: assembly-line production in consumer goods industries in the period between the two world wars. She argues that the new mass production of consumer commodities depended on women not only as a source of competent and cheap labour at work but also as purchasers and users. They were the necessary consumers of the very commodities they produced at work (Glucksman, 1990).

These feminist historical studies combine to highlight a number of important factors missing from mainstream social studies of technology. First, the focus on women brings more clearly to view the meaning of the link between production and consumption, production and reproduction. Second, questions are raised as to *who* designs, *who* sells and *who* uses. Does their gender matter? Third, there is an emphasis on culture that is lacking in mainstream social studies of technology. Michele Martin writes of 'the culture of the telephone', and Louise Walden defined culture as crucial for her project. She describes the ruling male culture in the male-dominated sewing machine industry and business in interaction with the female culture of unpaid and paid sewing in the home (Martin, 1991; Walden, 1990).

Fourth, these studies show that technological change is quite capable of transforming detailed tasks and activities without changing the fundamental asymmetry and inequality of the relation between women and men. As Louise Walden says, 'In both men's world and women's world, the sewing machine affected the work. Both the male and the female culture had to adapt . . . but the *relation* between male and female culture remained unchanged' (Walden, 1990: 263). Michele Martin found that the telephone offered women a new means of communication, yet it 'did not have a revolutionary effect at the level of gender relations. It tended, rather, to reproduce the patterns of male domination' (Martin, 1991: 171). Cynthia Cockburn's study of the introduction of electronic technologies into contrasted work processes found a similar persistence of gender relations from one technological regime to the next. Despite a change in the pattern of jobs, it continued to be men who gained the technological knowledge and know-how that permitted control of the equipment, while women's understanding was limited to the little needed for its straightforward operation (Cockburn, 1983, 1985).

Finally, all these studies are expressive of a feminist political project. There is a persistent feminist concern that by virtue of being the inventors and designers of technologies that women use, men enhance their domination of women. As Ruth Schwartz Cowan says, 'tools are not passive instruments, confined to doing our bidding, but have a life of their own . . . People use tools to do work, but tools also define and constrain the ways in which it is possible and likely that people will behave' (Cowan, 1989: 9). Being tool-makers to the modern world, men control the means of women's labour (Cockburn, 1985). This concern over the implication of technology in the control of women has led other feminists to make a critique of technologies that have particularly far-reaching effects on women, women's work and reproduction: nuclear weapons and nuclear power, for instance (Bertell, 1985); developmental agricultural techniques in the Third World (Shiva, 1989); or reproductive technologies (Stanworth, 1987). Such studies identify certain feminine cultural spheres – nutrition and horticulture, contraception and childbirth – and argue that innovations produced within the white, male, Western monoculture of technoscience have a damaging effect on women's interests and indeed are dangerous for all concerned. Sally Hacker went further and suggested that in the masculine culture of engineering, technology, eroticism and power are linked (Hacker, 1989), and Janine Morgall argues the case for a specifically feminist practice of technology assessment (Morgall, 1991).

Our microwave study takes its inspiration from these antecedents, picking up and developing many of the concepts noted above. It is, we hope, new in applying methodology and concepts from mainstream social studies of technology where these seem productive, while making gender central to the analysis and feminist change the purpose of the project.

The Organization of the Book

The sequence of chapters is as follows. In Chapter 1 we tell the story of the development of the microwave oven and its associated cooking practice. We sketch the circuit of its conception and design, development and manufacture, marketing, retailing, use and servicing and show how this becomes visible at closer range as a complex actor-world. At the heart of this microwave-world are the large organizations that on the one hand produce and on the other distribute microwave ovens. Inside them many local actors can be distinguished – units, departments, individuals. Outside, yet more interested parties are involved: advertising agencies, women's magazines, groups concerned with public safety and most importantly the customers who do (and do not) buy and use microwave ovens.

In Chapter 2 we ask, 'Where are women and men located in the microwave-world?' We identify the gender pattern of people's location, the different positioning of women and men in the various sites and activities producing the new cooking. Women are present here in distinctive roles: as people whose 'nimble fingers' are useful in production, as home economists whose cooking knowledge is needed in design, as microwave consultants and sales assistants who can relate to women customers, as wives and mothers who cook for their families. Men have two particularly influential roles: as managers and as engineers. They are the technically informed actors. We note the inequalities that ensue from these marked sexual divisions: unequal pay, unequal opportunities for training and promotion, more and less mobility, leisure, and initiative.

The difference in where and with what effect women and men are practically located in relation to technology is only the beginning of the story. Occupational sex segregation is nowhere total, and women in particular increasingly step out of line. Locations and activities can change their gender associations. Yet – somehow – gendering and gender hierarchy persist. The following three chapters show something of how this happens, in the interplay of gender identity, gender structure and gender symbolism; between individuals, the gender pattern of location and the deployment of meanings. We look at technology and gender as represented in culture, in artifacts, words and images. We show how the masculine and the feminine are continually reasserted as complementary spheres, the former with higher status, greater value and authority: simply more important.

Chapter 3, then, draws on material from Electro, the Japanese company producing microwave ovens. The all-male teams of product planners, design and production engineers in Japan and the UK lead the development of a line of microwave ovens. An all-female group of home economists are employed to 'stand for' the presumed female user, to input cooking knowledge, testing the various models and designing instructions and recipes. We see how engineering and home economics are ascribed different genders, contrasted meanings and unequal values.

In Chapter 4 we move on to the second major site in the microwave-world: the retail 'multiples' that sell microwave ovens along with other household electrical goods. Here we see the way the sale and purchase of domestic products is split on gender lines. 'Brown goods' (music systems, TV, video recorders) are represented as interesting, important and masculine; 'white goods' (fridges and freezers, washing machines, vacuum cleaners) as 'family' goods and as technically boring. The microwave oven begins its life with an ambiguous identity – an almost-brown good with interesting technological features – but eventually slips firmly into line down among the white goods. Contradictions, however, are increasingly occurring here as retailers see economic advantage today in enrolling women as well as men in selling technological products, men as well as women in buying domestic equipment.

The narrative moves in Chapter 5 to the third main site of the microwave-world: households using microwave ovens. The microwave oven is clearly a welcome innovation in the homes of Britain, but it is not always used in the way manufacturers and retailers intend. Women and men are changing their cooking and eating habits. The microwave oven is implicated in the change. But we find this new technology, like so many before it, shaped by the gender relations it enters – in particular those of the heterosexual domestic couple.

The knowledge through which technology is shaped and used, the processes in which it gains its significance, even the artifact itself, are moulded by gender relations. But such technology relations in which we are all, one way or another, caught up, are also among the processes in which gender difference, complementarity and asymmetry are adaptively reproduced over time. Doing, making and producing, constructing or using technologies, are a daily context in which each of us deals creatively with our material circumstances, and the representations of gender that reach us, to evolve our individual subjectivities.

In the final chapter we shift the emphasis from the shaping of technology to the shaping of gender identity in and through the technology relations of this microwave-world. We show how men, masculinity and technology-as-engineering are ascribed relative importance; women, femininity and other, particularly domestic, technologies are given a lesser value. We draw out the ways in which, through the technology/gender relation, gendered subjectivities diverge, with men developing a greater sense of agency, women of sustenance – and consider a feminist strategy of change.

1
Achieving a New Technology

The microwave oven: it sits there, in the kitchen, in the advertisements, white, rectilinear, trimly styled, somewhat enigmatic, banal. Fifteen years ago cooking by microwave radiation was a novelty. Today it has its accepted place in the routines of more than half the households in the UK.

Yet the enthusiasm with which the public took the microwave oven into its kitchens was a particularly British phenomenon. By 1984 the UK had almost 2.5 million microwaves in use against less than 1.5 million in all the other countries of Western Europe and Scandinavia combined. The number increased fourfold in the prosperous years 1984–88 when the value of UK annual sales topped £400 million. More than 1.8 million microwaves were sold in the peak year of 1988 alone. Though the Scandinavians eventually overtook the UK in terms of microwave ownership, in 1992 the UK still leads the remainder of Europe with penetration of between 50 and 60 per cent of households (estimates vary; see Marketpower Ltd 1989, 1991; British Market Research Bureau, 1990). The rapid growth of microwave sales in the mid-1980s was powered by first-time buyers. The great marketing advantage of the microwave oven is that it is an *addition* to our domestic equipment. It does not imply, as many new technologies do, the scrapping of a preceding one. To sell a microwave, the market does not have to wait for the formation of a new household, or for a conventional oven to become obsolete. Conversely, it is subject to great elasticity of demand, sensitive to levels of prosperity and to any government 'credit squeeze' (Economist Intelligence Unit, 1990).

Nonetheless, by 1989 the artifact's relationship with the public was maturing: 10 per cent of sales in 1989 were replacements for earlier purchases. A good demonstration of the still innovatory character of the microwave as a domestic artifact is to compare it with another relative newcomer to the kitchen: the freezer. The freezer boomed earlier than the microwave. In 1985 it had penetrated 36 per cent of households against the microwave's 10 per cent. By 1989 the two were neck and neck at 40 per cent but by then more than half all freezer sales were replacements for old freezers (Keynote Report, 1990).

Those one in two households in the UK today that possess a microwave, or are even on to their second or third model, appear to span all classes, age groups and lifestyles. There is, however, a slight weighting towards the skilled manual class and to 25–44-year-olds, the age group most likely to be establishing permanent homes and having children (Keynote Report, 1990).

What is so different about microwave cooking? We are used to thinking of the 'new technology' of the second half of the twentieth century as meaning the computer, its electronic components, the silicon chip. But there are many other significant new technologies transforming life and work today: technologies of light, like lasers and fibre optics; genetic engineering and other interventions in human reproduction; nuclear fission and fusion. The microwave oven too in its small way is 'revolutionary'. Through all the millennia of human history we have cooked food by applying fire – directly as in roasting, indirectly as in steaming and boiling. A microwave oven is entirely innovatory: it cooks food by bombarding it with radio waves so that it in effect cooks itself. The radiation causes absorptive particles in the food, particularly ions of salt and water molecules, to reverse their polarity about 5000 million times a second. So energetic is their movement that the friction generates intense heat, so bringing about the changes in food we call 'cooking'. In fact the microwave radiation is able to penetrate no more than 3 or 4 cm. below the surface of the food. Heating of more solid and deeper parts of the food-mass occurs by contact with adjacent irradiated particles.

The oven that produces this novel effect is a rather simple artifact. It consists of a box with an outer metal casing and a reflective inner cooking chamber. It has a well-sealed door, with a glass panel for viewing. Its most important component is a magnetron which converts household electric current to high frequency (2450 MHz) electromagnetic waves. A system of guides directs the waves into the inner oven and a fan distributes them evenly over the food. A turntable rotates the food container. Since microwave energy penetrates glass and plastic, food containers used in microwave cookery are usually made of these materials. Later models of microwaves have electronic controls that govern power output, program the timer and respond to the 'doneness' of the food, and it is these that constitute the most complex aspect of the machine.

The attraction of the microwave is, first and foremost, speed of cooking. A baked potato that might take forty minutes in a conventional oven will cook in four in a microwave. Less forethought is called for: the oven requires no pre-heating; frozen foods can be defrosted and cooked in one sequence. Effort is also saved in cleaning. The oven itself gets less soiled, and washing up is reduced since food can often be cooked and served in the same container. Small portions can be prepared without waste of energy. Leftovers can be reheated without deterioration. Individual tastes and time schedules can be catered for. There is an environmental benefit too: unlike a conventional cooker, the microwave oven does not heat up the cooker itself, the kitchen and the external atmosphere.

There are disadvantages to microwave cooking, however. Although some foods cook better in a microwave, certain nutrients are retained better and there is no transfer of flavour when several items are cooked together, not all foods respond well. Pastries and roasts are not successful. Large items and deep dishes of food do not easily cook to the core. Accidents can happen. Experienced cooks and perceptive eaters have these and many other complaints which we shall hear in the course of this book. Nonetheless the microwave oven has found its niche in our lives and seems to be here to stay.

Like many new technologies introduced into factory, office and home, the magnetron was first developed for military use. It was in 1940 that scientists in the Department of Physics at the University of Birmingham first made and operated a cavity magnetron with the potential for development into the microwave oven as we know it today. The interest it evoked had, at this stage, nothing to do with either heat or food. As subsequently manufactured by GEC the magnetron gave Britain an early lead in radar. It enabled radar equipment to be built that was smaller, more powerful and more accurate than anything previously designed. Britain, then at war with Germany, sold its magnetron know-how to America in exchange for mass-production facilities.

It was not until 1945 that scientists in the US radar technology firm of Raytheon Ltd woke up to the magnetron's heating capabilities, patented a device, and some years later approached the Tappan Stove Company to manufacture under licence the first microwave oven. Soon similar licences had been agreed with production companies in the UK, Germany and Japan. The first ovens produced on any scale were catering, rather than domestic, ovens. In the 1970s, however, the market for domestic ovens began to expand and many new companies, American, Japanese, British and others, joined the pioneers. By 1980 the world market for microwaves had reached 4 million per annum (Andrews, 1990).

From Conception to Production

The use of the magnetron in military radar in the 1940s was achieved by the cooperation of an array of actors (for an explanation of our use of this term see p. 10). These included Boot, Randall and Sayer, the three Birmingham University physicists; the celebrated scientists Sir Henry Tizard and Sir John Douglas Cockroft; even Winston Churchill and Harry Truman. The magnetron as a commercially applicable invention turned out, as we have seen, to have what is termed in social studies of technology 'multidirectionality'. It could be turned to ploughshares as well as swords, baked potatoes as well as the detection of enemy aircraft. Converting the magnetron to its new use in the 1950s and 1960s and launching it into the world's kitchens in the seventies was achieved through the association of a second cluster of actors: Dr Percy Spencer of Raytheon; the risk-taking directors of the Tappan Stove Company; and some ambitious strategists among the manufacturers of electrical consumer goods operating on a world scale.

In this chapter we will trace the relationships of their inheritors, the contemporary actors through whom (and despite whom) a range of microwave ovens, initiated by a manufacturing firm, takes its ever-evolving shape, reaches its public and modifies their cooking practice.

Note that it is 'microwave cooking' we call the innovation we are studying, not 'the microwave oven' (see Figure 1.1). This is a reminder that technology is not just hardware. As Donald MacKenzie and Judy Wajcman put it '"technology" refers to human activities as well as to objects'. It also 'refers to what people *know* as well as what they *do* . . . Technological "things" are meaningless without the know-how to use them, repair them and make them' (1985:3). So our questions will be about more than the artifact. What is the cast of the microwave drama? Who are the actors who have (with moderate, limited success) changed the nation's cooking and eating behaviour? How did they do it and in what relation to each other? Who was pushing, who was pulling? What were the materials, and the material constraints, they had to deal with? We will ask all this in order to press on to the further question: where are women, men and gender relations in the microwave-world? For the microwave oven is of little societal interest so long as it gathers dust on the retailer's shelf or clutters up the kitchen worktop, unused.

En route to finding out about the technological actors, knowledge and processes, we will not of course neglect the microwave-as-white-box. Its stand-alone simplicity, however, is even then misleading. What comes to the user is in fact a package of products. As can be seen from Figure 1.6, ahead of the microwave oven comes its promotional image, the way we are invited to see it. In addition, with it come instructional materials, carefully written and illustrated to ease the changes in cooking practice the user must adopt. Only by such accompaniments will large numbers of people consent to participate in the microwave cooking project as customers who in turn become users (cooks and eaters), so establishing this innovatory cooking practice in popular culture.

In the early days of microwave use in the UK most of the ovens were imports from abroad. However, UK production grew by a factor of nine between 1984 and 1988. Much of this production was in turn exported to Europe and elsewhere. Yet imports did not rise greatly over this period, indicating that the waxing enthusiasm of the British householder was increasingly being met from local production (Association of Manufacturers of Domestic Electrical Appliances, 1991). One of the new manufacturers in the UK was the Japanese multinational we will call Electro.

Electro, founded early in the century, made its reputation in communications equipment. It began manufacture of microwave ovens in Japan

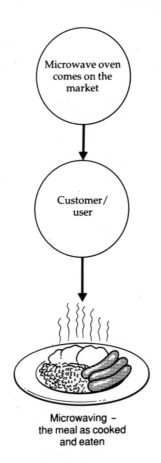

Figure 1.1   *An artifact becomes a technological innovation only when used.*

in 1962 and developed a growing export market. Today, as one of the several Japanese giants involved in production of leisure electrical goods, office equipment and household appliances, it operates dozens of plants around the world and employs many thousands of people.

The firm quickly developed a big name in microwave ovens in the UK. Like its giant Japanese competitors in the electrical products field it has recently followed up its high import penetration by establishing manufacturing plant in the UK, and responding to anti-dumping moves by the European Community in anticipation of the Single Market in 1992. Japanese electronic companies' investment in the UK had reached £1324 million by 1990 (EIAJ, 1990).

Undoubtedly an added attraction to set up locally was the availability of relatively cheap land on a green field site on the Celtic fringe of Britain, situated in a travel-to-work area in which traditional industries

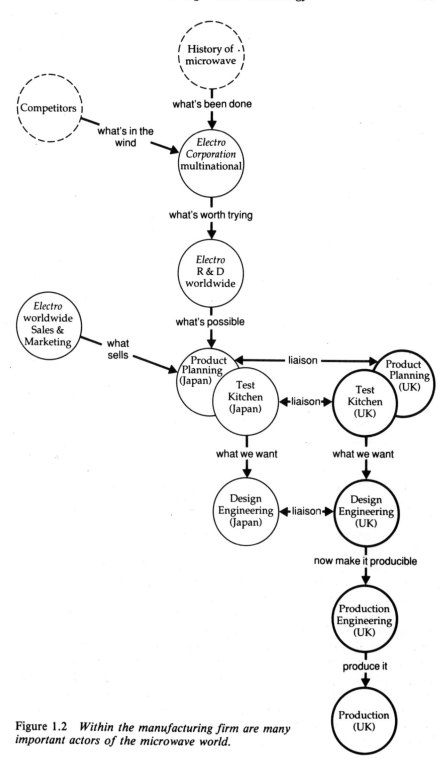

Figure 1.2 *Within the manufacturing firm are many important actors of the microwave world.*

were in profound decline, resulting in around 20 per cent unemployment. A Labour Council, delighted to obtain the promised jobs, brokered a 'single union' deal with the Electrical Electronic Telecommunications and Plumbing Union (EETPU). It appears that a second term of Thatcher government had reduced workers on the British periphery to near Third World status: the labour force that Electro stood to acquire could be guaranteed to be experienced, disciplined, docile and cheap. The new factory in the countryside began production of microwaves and business equipment in 1986. By 1991 Electro had become one of the UK's largest 500 companies (Allen, 1990: 42). It had invested over £52 million in the UK, including a head office in a major city, a research and development centre and an outlying production plant, and by now it employed 1200 people.

Our sketch of the contemporary microwave-world then begins with this key actor: the Electro Corporation. As we examine it, however, Electro quickly fragments into a number of differentiated actors, each with its part to play. Figure 1.2 shows on the left Electro Japan and on the right Electro UK, their internal structures to some extent analogous. (Electro of course has many other subsidiaries strategically sited in other regional markets, in a similar relation to Japan.)

Electro Corporation has a long history of fielding its own range of microwave ovens. It has learned much from the past successes and failures of its microwaves and those of its major competitors – mostly US and Japanese. The firm has a large research and development department at head office in Japan and R&D centres in various countries around the world, including the UK. They are in continuous interaction and none works exclusively for its local market. In Japan there exists a sub-section of R&D for microwaves, but teams concerned with, for instance, materials testing and electronic components have an input to microwave design too.

The heart of the microwave project is Product Planning, centred in Japan but also featuring importantly in the UK firm. Until around 1990 all Electro products were designed and developed in Japan for the world market. 'The basic design constraints came from Japan and we were left to do the styling,' said one British engineer. Classic misjudgments occurred due to ignorance of local cultural preferences. The recollection of a pink-and-blue microwave still raises smiles in the UK planning department – smiles of embarrassment from the Japanese engineers, 'told-you-so' smiles from the British. Today Japan is more concerned to take account of locally generated market knowledge. The catch-phrase is 'global localization'. Cultural specificity is particularly important in the case of a cooking technology. Italians like to cook *au gratin*. The British like to roast. The Japanese experience the particular need of defrosting raw fish, for *sushi*, without permitting it to cook. Such lessons have been learned by microwave manufacturers the hard way. So today the UK product planning team, instead of being obliged to pick a range of models off the Japanese shelf, can increasingly negotiate modifications or even specify and design their own models for the local market. This compromise accords well with the Japanese management style which favours group decision-making. 'No one man's opinion necessarily wins through,' we were told by the Microwave Product Manager. 'It's debated among all types of managers . . . Always this pooling of resources and information.'

Just as it is inappropriate to consider the microwave as artifact, without also considering the image that is projected of it and the instructions that accompany it, so it is best to see any one model as part of the manufacturer's current *range*. Within the frame of multidirectional possibilities the manufacturer keeps some options open. What the Product Planning Department have in mind in specifying any one microwave oven is in fact its place in a *set*, a set of price points, cubic capacities, power levels, colours, etc., an Electro range of microwaves that will as a whole satisfy all the needs of the European customer. Or, rather, those on which Electro Corporation wishes to compete with other manufacturers. For the competition too is a shaper of Electro's range. While we were undertaking this research a Moulinex advert began to appear on French hoardings launching a high-power 1100 watt microwave. The theme of the ad was 'watch for the competitors' response'. Ripe tomatoes were splashed across the image. Some manufacturers would no doubt respond to this challenge by increasing their own maximum power rating. Others would scorn it as a misguided strategy, and instead strengthen the reliability and controllability of their own top-of-the-range models.

We focus now on Electro UK (see Figure 1.3). In making its decisions, UK Product Planning draws on data from both within and outside the company. Internally, information reaches them from Electro's Sales Department, via Marketing, on the relative success of existing models. Customer Relations has more precise advice to give based on customer complaints. The product planners also have access to a range of external sources, including the purpose-designed market research carried out by the advertising agency retained by Electro, national surveys of retail sales, and major consumer research exercises such as the Target Group Index and National Buying Survey of the British Market Research Bureau.

In the heart of the Product Planning Department itself is a further source of advice and information: the Test Kitchen. As we shall see, the Test Kitchen mainly looks in the direction of Design Engineering, testing prototypes and writing cooking instructions. However, it embodies cooking knowledge for the whole company and its staff are, since they actually do cooking, of all the Electro employees, closest to the customer/user/cook.

Product Planning and Design Engineering work interactively both within Japan and between Japan and the wider corporate system. The

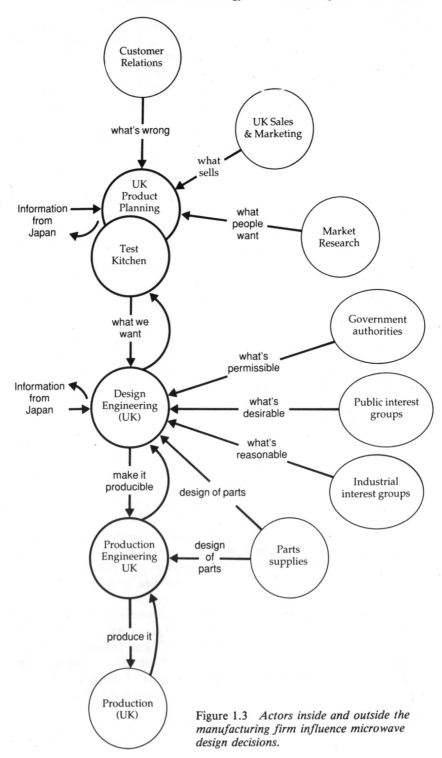

Figure 1.3 *Actors inside and outside the manufacturing firm influence microwave design decisions.*

two UK departments, together with the Quality Control Department, have monthly 'design review' meetings around new products. If the role of Product Planning is specification, that of Design Engineering is to design. This is a less glamorous role than might at first be thought, since a 'new' microwave oven is always an adaptation of one or more existing models. Ovens, and indeed the entire range, do not spring new-born from the head of the design engineer: they evolve. Besides, the work of Design Engineering UK has in the past mainly been to modify drawings sent from Japan. At the time of our study they had recently, and for the first time, been given full responsibility for producing a new model for the European market.

Design Engineering too draws on a wide range of sources in addition to the know-how of its own engineers. Of course colleagues at Electro Corporation in Japan will be the first source of knowledge, but microwave specialists are a world-wide fraternity. Developments are in hand for applying microwave heating for many different purposes: curing rubber, for example. Knowledge of this kind may be transferable. Improvements are always being made to the understanding, for instance, of cavity design. The cavity of a microwave must be electrically matched to the magnetron. It must have appropriate resonance, impedance and security with regard to the electrical wavelength, so that it will enable all-round heating and prevent leakage of waves into the environment. In-house engineers will use CAD programs modelling the Maxwell equations, the relationship between magnetron output, wave guide and cavity materials and shape, developed in research laboratories elsewhere.

One aspect of design, and the one perhaps most obvious to the everyday user, is 'industrial' design: giving the oven its external style. Industrial designers are trained to design many different kinds of appliance, machine or utensil. A good designer will be able to turn a hand to shaping anything from a hair-drier to a photocopier or a wheelchair. A microwave is in many ways a rather simple industrial design project: it is almost essential that it be rectilinear and but for door, turntable and fan, it has few moving parts. Nonetheless it must be in keeping with current taste. 'The whole of our industry is a cosmetic industry,' said the Sales Manager. 'Every six months the fashions change.' It is the industrial designer's job to bring to the Design Engineering team a feel for what changes are occurring not only in the field of microwave ovens but in that of all kitchen technology, and indeed all consumer goods. Is it sharp angles people want today, or softly rounded curves to the corners of window and control panel? Which will sell best at the moment – pronounced or barely visible command icons, projecting or flush keys and knobs, recessed or protruding handles?

One input to industrial design must be interior design, for the microwave oven has to argue its way convincingly into the kitchen. There are two parallel tendencies in kitchen design, both claiming to advance

beyond the 'steamy, smelly workroom' of the past. One is often dominant over the other in popular appeal at any given moment, but both
continue to inform the advertising of kitchen equipment and complete
kitchen systems. One is the clean, white, sterile machine-for-cooking-in
typified by the Miele kitchen: 'Engineering. In a Miele kitchen everything
runs, rests or hinges on it.' Miele advertises its kitchen as feeling like an
expensive car. 'Smooth, beautifully finished surfaces conceal thoughtfully
conceived engineered details . . .' The emphasis is on efficiency and
work. The second and competing style is typified by Wilson & Glick's
'traditional country range', dark in tone, softly illuminated, conjuring up
'an image of harvest time, the smell of woodsmoke, drying herbs or
freshly baked bread'. The emphasis here is on relationship. This kitchen
is not a workshop but 'a room like other rooms', 'the heart of family
life'. The microwave oven in its early days contributed to and gained
from the 'engineered kitchen' movement. It still has no problem, with its
neat geometric appearance, finding its slot in the efficient kitchen. The
greater challenge for the industrial designer is creating a microwave that
does not stick out like a sore thumb among oak veneer panels and chintz
curtains.

The components that go to make up a microwave oven come from a
multitude of different suppliers. The magnetron itself is a special case,
coming as it does from one of only four manufacturers in the world.
Electro Corporation prefers to supply from Japan certain critical items,
including the door, the fit and seal of which are important for safety.
The internal cavity and external wrap, however, being essentially boxes
full of air, are too expensive to transport around the world. These and
many other components it is cheaper to source locally. Some will be of
very general applicability – for instance the transformers, capacitors and
diodes that make up the power source for the magnetron. Some, such as
the control panels, must be designed or specified by Electro in precise
detail. Though their deliveries are subject to rigorous vetting by Electro
UK's Quality Control Department (and some are even sent to Japan for
clearance), it will be evident that parts vendors themselves contribute
something to shaping the product, whether by supplying from stock of
their own design, or by producing custom-made components.

The Test Kitchen, though located in Product Planning, is in many
ways a working part of Design Engineering. A new model of microwave
oven is performance tested on four occasions at different moments of its
evolution. First, the Test Kitchen receives a hand-made sample from
Design Engineering. Later, they get an off-tool sample, one of a small
batch produced under Design control. Third, they test a production
sample from a run of around 100 ovens organized by Production
Engineering. And finally, a random sample is withdrawn from an early
mass production run.

The home economists that staff the Test Kitchen put the microwave
through its paces. First they apply certain basic performance tests to see

that the artifact is up to international and national safety standards. Second, they move on to company tests, appropriate for the particular model, using constants supplied by Japan. Under various rigorously specified conditions and with strictly measured quantities, shapes and densities, the oven is called upon to heat many loads, from water to mashed potato, egg-custard to meat loaf. The results are measured, weighed, colour-tested, stirred, licked and sniffed. Power output, oven heat and the performance of all the various forms of sensor equipment – the steam sensors, temperature probes, weighing scales designed into some microwave ovens – are recorded. Results are fed back to Design Engineering in the UK and also to Japan. 'Because they are, and they like it seen that they are, the bosses,' said a home economist.

The Test Kitchen often advises Design on improvements that could be made to control panels, and provides the data from which software will be written for the programmable models in the oven range. A further important task of the Kitchen and Design Engineering together is to devise an interface between the oven and its eventual retailers, installers and users. The Test Kitchen write both elaborate recipe books to promote microwave cooking generally, and precise cooking instruction booklets to accompany the oven out of the factory: how to keep your oven clean, what power levels are appropriate for what food, what kind of containers are and are not suitable for microwave use, how food must be stirred or turned while cooking, or left to stand before eating. Simple example recipes are usually given in these booklets. The instruction material guides the retailer in selling to and advising the customer, and it is important in ensuring the user does not get bad (or even dangerous) results from microwave cooking that could bring Electro ovens and the whole microwave cooking process into disrepute with the public.

Design Engineering, for its part, take responsibility for a further two aspects of design: adherence to safety standards and the interface between manufacturer and installer. On the first count a Safety Standards Engineer is employed who involves officials of standard-setting authorities in approving designs and prototypes. On the second, it prepares technical data on the oven presented in simple terms so that retailer, electrician or householder will know the technical specifications of the artifact, will know what power point to use, what fuse to put in the plug, in what environment the oven is best installed and so on. Industrial designers will add to the effectiveness of this carefully crafted interface by supplying informative peel-off banners to stick across the corner of the oven door or on the control panels. All of these co-products of the artifact are important in facilitating its transition from one of ten thousand factory products to one in a thousand retail commodities in a high street shop and eventually to the place that really counts for the success of the whole project: a secure position as some household's unique tool in daily use.

A new prototype oven then is not the brainchild of Design Engineering alone. There are inputs from many actors in what becomes the microwave-world, some internal to Electro, some external. Above all, the prototype is produced and modified in an iterative, negotiated, process between Design Engineering and Production Engineering. It is rarely obvious to the outsider that design is only partly a process of evolving the artifact that has optimal qualities in *use*. As important to the manufacturer is that it should be optimally *producible*, with a given workforce, plant and factory, at a competitive cost. Production Engineering notionally takes the artifact to pieces in order to work out how it may be built again on a continuous flow assembly line by relatively unskilled workers in long, trouble-free production runs.

Even the people of the Celtic hills in their low-paid assembly jobs are actors in the shaping of the microwave oven. If their hands and fingers rebel at certain tasks on the line, and if automation is not a viable alternative, the build must be modified to accommodate them. In a more general sense, the economic circumstances and organized strength of the labour force in a given country or region will determine for the company the degree to which it is advisable to automate production. This too will partly shape design.

Design Engineering's special knowledge concerns the technology of the product. That of Production Engineering concerns process technologies. These engineers specify (and Design Engineering approves) the tooling for the production line. They allocate labour, calibrate equipment and issue tools. As the new microwave oven passes from the hands of Design Engineering to those of Production Engineering and thence to Production itself, novelty must sometimes be traded off against production convenience, or appearance against the cheapness or availability of certain materials. Changes are being made up to the last moment. Only when a first prototype, then an off-tool batch, then a trial production run have all been deemed successful – satisfying Design Engineering as to product, Production Engineering as to the 'balancing' of the line – will the assembly belt roll and the artifact begin to find its way out of the factory, safety-tested, quality-checked, shiny and inviting, neatly packed in polystyrene and cardboard.

Product Becomes Commodity

The finished microwaves move out now as commodities into the hands of other actors: the warehouses and shops of the distributive system. All the actors in Electro, diverse though their roles might be, and though some will carry the torch for one innovation and some for another, are all united behind the microwave cooking project. There can be no guarantee, however, that the retailers share their enthusiasm. Their main concern is to gain the maximum turnover of stock. Every commodity must earn its floor space: money per square foot is what counts. The retail trade must

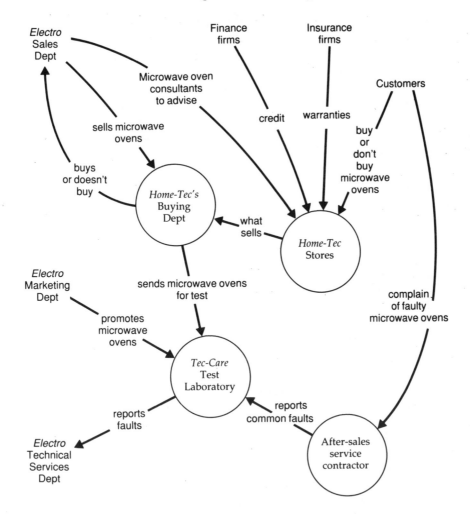

Figure 1.4   *The microwave actor-world includes the retail firm and its many associates.*

therefore be actively wooed into cooperation with the manufacturer concerning the microwave project. This is achieved by the manufacturer by a variety of strategies.

First, the product itself has to have what it takes. It must have winning looks, reliable performance and be competitive on features and on price. It must also be backed by a hassle-free manufacturer's guarantee. Secondly, the Electro Sales Department maintains close and engaging links with the retailer, giving advance information on new models or price changes, advice on selling points, eye-catching point of sale promotional material, intelligent advertising and agreements to share advertising costs.

Thirdly, along with its sales staff, Electro UK fields a team of microwave consultants ready to go out into the shops, demonstrate the features of Electro microwave ovens and answer the queries both of retail sales staff and customers. Since a manufacturer benefits from an effective retail sales staff, Electro encourages its consultants and sales team to put some effort into improving the product knowledge of the retailer's personnel. It has, for instance, sponsored a retailer's training video on how to sell microwaves. Some of these relationships are illustrated in Figure 1.4.

In the early days of microwaving, the ovens were commonly sold by small, independent, specialist shops. The larger retailers were slow to be convinced. When they did take up the microwave, however, they quickly overwhelmed the small companies, many of whom went out of business. At the same time, consolidation was occurring among the big electric retail firms. We carried out a detailed case study of one of the most important of the resulting chains in the electrical goods field – one we will call Home-Tec. The firm is typical enough of others of its kind, operating several hundred outlets, some of which are high street shops, others edge-of-town superstores. We also interviewed in one of Home-Tec's several competitor chains, which we call Wonderworld. We shall get a closer picture of both companies in Chapter 4.

Home-Tec had its own strategy for engaging with the manufacturer. The senior buyer responsible for microwave sourcing maintained energetic contact with an opposite number on the manufacturer's sales team. Slow-moving models cluttering the retail shelves are a source of complaint. New models are scrutinized and compared with competitors' ranges. What the buyer requires is an appropriate range from all manufacturers combined. If a certain price point or feature cannot be obtained from one manufacturer it will be sought from another. In this sense Home-Tec and other retailers contribute to the function of competition in the market. The bigger retailers commission microwave ovens for themselves, buying direct in bulk from manufacturers prepared to see them branded and sold under the retail chain's own name. These 'own brands' are often in the lower price bracket. A manufacturer like Electro therefore stands little chance of competing on price at the cheapest end of the range and must hope to obtain orders from the big chains for the more complex and costly models. Electricity Board shops, on the other hand, another extensive retail outlet, do not have their own brand and will take anything that sells well.

The retailer's microwave buyer has access to independent reports on many models. Information on sales performance is regularly returned by the stores in the Home-Tec chain, which have an efficient Electronic Point-of-Sale (EPOS) database. It is quickly known which artifacts appeal to the customer and which do not. The retailer also plays a part in design by feeding back to the manufacturer highly specific sales information to supplement that obtained from market research firms. One retail buyer told us the great challenge for me is not only the selection of the products that exist, but influencing the design of products. It's the buyer's responsibility, especially when you have 20 per cent of the market, not to allow a product to come on to the market that shouldn't be on the market.

Once sold, the microwaves still generate useful information that contributes ultimately to design. All customer queries and complaints are logged on Home-Tec's database. Though 'closure' has already occurred on current models, it is not too late to use the possibilities of multidirectionality, to inform the design of the range as an ongoing project and shape the microwave cooking process in historical perspective. Microwaves characteristically come with a one-year manufacturer's guarantee, but the well-trained sales person attempts to persuade the customer to add an optional two- to five-year extended warranty. Home-Tec subcontracts after-sales servicing. A firm we call Tec-Care plc handles its brown goods servicing, while white goods are dealt with by Nationwide After-Care plc. If the oven is still under manufacturer's guarantee it may be returned to the plant; alternatively the subcontractor may do the repair and recover the cost. After-Care perform a monthly fault analysis, brand by brand. Microwave ovens, out there at last in the household, send back messages. Those that lie quiet are good news. Those that are returned with factory faults or often require the call-out of a service engineer flash warning lights to the retailer's buyer.

Home-Tec can minimize the chance of trouble by commissioning its own technical tests to products it is considering stocking or with which trouble is being reported. Thus Tec-Care serves also as Home-Tec's test laboratory where examples are taken to pieces, components checked, sometimes tested to destruction. 'We're paid vandals really,' said a manager. They not only scrutinize the artifact, they also check the instruction booklets to ensure that the information they communicate is complete and understandable to the customer. All customer enquiries are put on a database and analysed. Shortcomings will be discussed with the manufacturer. Electro UK has its own Technical Services Department that is in constant touch with such retail test-beds.

It will be evident that two outlying kinds of organization here become part of the social world that generates microwave cooking. One is the insurance companies that underwrite the extended warranties. Their actuarial calculations concern the quality of the artifact and its tendency to failure. The second kind is the finance companies that specialize in hire purchase arrangements for the purchase of consumer durables. Their calculations concern not the trustworthiness of the artifact but the creditworthiness of the customer. Between the two the onward movement of the artifact into the home is eased. Their importance should not be underestimated. As we shall see, cheapness and above all cheap credit, 'terms', is the main selling point in the case of household equipment. And warranties are such a gold mine to Home-Tec and other such firms that they currently generate a substantial proportion of total profits.

We begin to see then how the achievement of a new cooking process is more than the design of a cooking tool. If they are to succeed, the artifact's originators must create a durable alliance with other actors. These have their own interests in the project. To a degree they may adapt or change the artifact, its image or its use, as the price of their adherence – they may exploit the technology's 'interpretive flexibility'. Some actors, the retailers for example, may in principle be enthusiastic since they stand to profit by any innovation that produces an item with rapid turnover. But they are fickle, and will quickly reduce an artifact's shelf space if a rival innovation seems to promise faster sales.

Advocates and Sceptics

Others whom the manufacturing/retail alliance needs on its side may be outright sceptical. Figure 1.5 shows a curious area in the microwave-world made up of parties with mediating roles to play between the manufacturer and the customer. What these actors purvey is less the artifact than that almost equally important product: its image, what people perceive it to be.

Manufacture of electrical goods such as microwave ovens is subject to official safety and quality standards set by international and national bodies. The standards of the International Electro-Technical Commission in Geneva and the British Standards Institute in London are part of the data input to the original design process. These bodies and others supply the minimum norms applied by the retailers' test laboratories. Such standards are of particular interest to one rather noisy actor, the Consumers' Association, watchdog for the customer/user. The CA, a voluntary body supported by individual and institutional membership, publishes a journal, *Which?*, in whose pages appear from time to time reports on microwave ovens. They arise from the CA's own lab testing of current models of microwave on the market, and from their surveys of consumer experience, which includes group discussions. They establish assessment criteria, and they rank and criticize individual models. Manufacturers sometimes loathe the CA, but they have learned to respect their reports and to seek a good reputation in *Which?*

Microwave ovens have been particularly vulnerable to challenges concerning their safety. The establishing of microwaving as a widespread habit, its normalization as a daily cooking practice, has been interrupted and threatened more than once by popular panics, which the media have reflected (or fomented, according to point of view). The first scare story rudely interrupted the honeymoon period of the public's relationship with its microwave. In 1979 a US TV programme was broadcast in Britain, raising fears of radiation leakage. The fears were proven largely groundless, but more rigorous factory testing of radiation security and improved design of oven doors resulted. Other scares followed in the

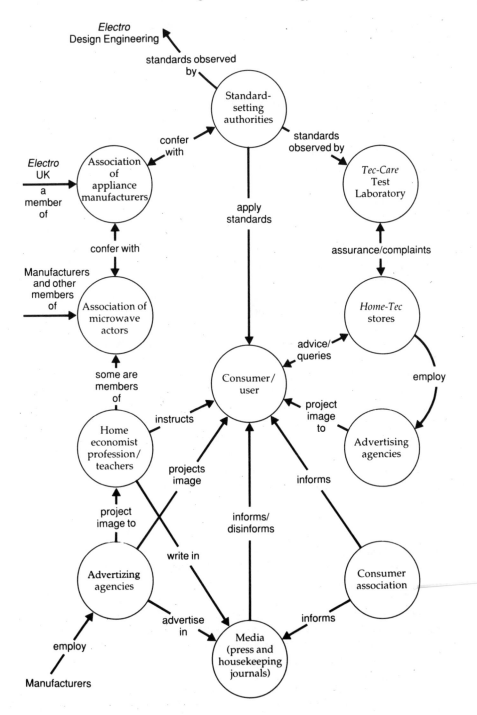

Figure 1.5 *Many institutional actors shape the consumers' knowledge about technology.*

1980s. Frequently a newspaper or radio or TV programme would carry a story of the explosion or incineration of food being heated in a microwave. Christmas puddings and mince pies, fatty, sugary and moist, are notoriously prone to over-cooking and fire in a microwave. Since both are festival foods served at a time of year especially sensitive for the public, it is not surprising if the media exploit such stories. The manufacturers and retailers, however, have begun to believe in some kind of conspiracy, in which the enemies of the microwave cooking project emerge annually around November to spoil pre-Christmas sales. A particular professor with a supposed grudge is occasionally named as the bad fairy. The reaction of responsible commentators is to refer users back to their instruction booklets. This, as we shall see, is problematic.

More recently, clingfilm, often used to cover food while microwaving it, was found to contain harmful plasticizers that could, under microwaving conditions, migrate into the food. The guilty plasticizer was banned and people soon continued with their usual ways. Undoubtedly such scares cause sales of microwaves to dip briefly and perhaps they relegate some ovens, already in the home, to non-use for a week or two. One incident however had a more serious effect: the poisoning scare of 1989.

In August 1989 the UK Ministry of Agriculture, Fisheries and Food (MAFF), published a report on tests of 104 microwave ovens which suggested that some constituted a health and safety problem. The cavity and wave guide of certain ovens were designed in such a way as to permit 'cold spots'. Some parts of loads not rotated or stirred during cooking would remain at too low a temperature to sterilize the food, while the remainder reached boiling point. Listeria bacteria could survive and poison the consumer. The news reached a public already alarmed by a wave of salmonella poisoning from inadequately cooked chicken and egg products.

The Ministry of Agriculture did not name the manufacturers of the faulty ovens since it had foolishly allowed itself to become indebted to them by accepting gratis, instead of buying outright, the sample ovens. The Parliamentary Agricultural Committee castigated the Ministry for irresponsibility. The Consumers' Association took up the alarm and protested that the Ministry owed it to consumers to publish names. The media joined in the hue and cry. Home-Tec and other distributors found themselves in an ambivalent situation, siding with the public in their suspicion of some makes of microwave yet wishing to cooperate with the manufacturers to combat adverse publicity and falling sales. The Ministry re-ran the tests in the autumn of 1989 with ovens that by then the delinquent manufacturers had hastily improved.

Meanwhile, however, it was beginning to be widely appreciated that in addition to design faults there was a communication failure between the manufacturers of microwave ovens and those of cook-chill and frozen ready-packed recipe dishes. A food research institute, alert to this problem, found that inconsistencies were occurring in the food loads on which different actors estimated and tested power output. What was emerging twelve months after the original scare was a new voluntary agreement between the food and electrical industries (which, though clearly actors in one and the same microwave-world, seemed seldom to meet or confer) on a system of 'banding' that would be applied to instructions to consumers on all microwavable foods. Ovens' power output would be re-established with standardized tests, the international IEC 705 code would be universalized and ovens assigned to one of a series of bands according to their cooking power. The Consumers' Association removed its teeth from the trouser legs of the Ministry and the manufacturers and turned its attention to finding out whether the consumer would be intelligent enough to understand the proposed changes (they would – more or less).

But people were alarmed. Sales of microwave ovens fell away catastrophically for a couple of months in late 1989, adversely affecting the year's performance for most manufacturers. The incident occurred in a period of sharp recession in the economy as a whole. The UK market for electrical household appliances fell by 14 per cent between 1988 and 1989, that for the microwave oven, on account of its particular tribulations, by 30 per cent (Keynote Report, 1990). In response, Electro UK cut its production to 65 per cent of previous output, and national production of microwaves fell by more than half. Home-Tec cut prices and reduced stocks.

The blip was not fatal to microwaving. Not long after the scare had blown over, a CA survey found, we were told, that 'people's practices haven't changed that much. They're still favourable to microwaves.' The market for microwaves began to recover. The revival was assisted by other actors who rallied to the support of the wavering project. The Association of Manufacturers of Domestic Electrical Appliances (AMDEA), representing the manufacturers, intensified its lobbying. Also active in trying to regain momentum for the project was the Microwave Association, an interest group with membership in many of the major organizations of the actor-world: the manufacturers, food and container industries, the electricity industry, etc. The Microwave Association sees itself as performing an educational role. It works closely with the profession of home economists, those in industry, in catering, in the media. The home economics profession can be an important ally or a dangerous adversary for the microwave alliance. Of special importance are those home economists teaching the next generation on home economics courses in higher education. The Microwave Association attempts to enrol teachers and students by offering discounted microwaves to the schools and awarding prizes for students.

Among home economists, a second influential group of actors is the microwave cookery writers. A mere handful, they produce a continual stream of cookery books, so that on any bookshop's cookery shelves, microwaving is always assured of its few inches. They write articles month by month for the women's, housekeeping and culinary magazines, reassuring readers on safety and quality, reminding them of the micro-waving 'rules', continually attempting to extend the menu range microwave ovens are seen as 'good for'. Some are commissioned by manufacturers to write cookery books for them. Others keep a greater professional distance from any one manufacturer and evaluate dangers responsibly. On balance, a vocal group of home economists has done a positive public relations job for microwaving. It has been an ally with the influence to enrol many more.

One could ask, with such friends burnishing the image that goes ahead of the microwave, why does the manufacturer need an advertising budget? And indeed there are limits to what advertising can achieve. At the time of the listeria scare, by common consent the manufacturers agreed to keep a low profile with the public until the trouble blew over. Most ceased advertising microwave ovens altogether. The agency retained by Electro had already decided, even before the disturbing appearance in the microwave world of these cultures of listeria bacteria, that the company had become too closely identified with microwaves for its own good and had advised its masters to set about a campaign to boost the standing of Electro as brand and as corporation in the field of electrical consumer goods generally.

Responsibility for advertising microwaves is shared between the manufacturer (with the agency it retains) and the retailer (with its own). Though Home-Tec were currently advertising their stores, mainly in general terms, in TV commercials, it is a peculiarity of advertising among retailers that they seldom single out appliances for advertisement, nor do they go in for colourful hoardings and double spreads in glossy magazines. Instead they rely on promoting the retail chain and its commodities in drab listings of current bargains in national and local newspapers, emphasizing only competitiveness on price and the avail-ability of favourable terms. The graphic artists in retailers' merchandis-ing departments also produce point-of-sale material, cards and peel-off stickers highlighting 'unique selling points' of different models. We should not underestimate the role of formal manufacturer and retailer advertising in the microwave project – certainly in the early days of microwave the TV campaigns of Electro and other manufacturers helped generate the boom. But it has to be recognized that, as much or more, the image of the microwave that reaches the consumer is generated by other image-purveyors such as the advertisers of microwavable foods, the home economics profession and the Microwave Association.

Consuming the Microwave Oven, Consuming Microwaved Food

The actor that is most crucial of all to the success of the microwaving project is that hydra-headed creature, the public: customer, purchaser,

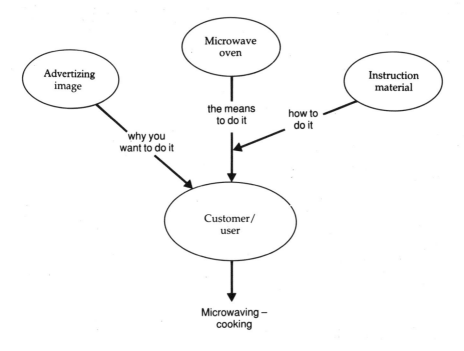

Figure 1.6 *The microwave received by the consumer is more than an artifact. It is also a projected image and controlled knowledge.*

user. When the individual or the household is given its full weight and presence in the microwave-world the only way in which to conceive the whole is as a sphere, in and beyond the volume of which are the many strands sketched in Figures 1.1 to 1.5. The consumers can then be seen as, so to speak, a spherical surface, in touch with other actors at many points.

As we have seen, 'dead' customers have influenced past sales achievements of earlier models of microwave oven and so contribute in their ghostly way to contemporary decision-making. 'Live' customers too vote with their purse. They read the newspapers and watch the commercials. They subscribe to *Which?* They discuss with each other and form opinions. Then they buy – or do not buy – a microwave oven. Unpopular models will soon be withdrawn or modified: 'back to the drawing board'. A sample of ordinary people is recruited to guide the decisions of the more institutional actors at the heart of the process: they are the ones who respond to market research questionnaires and consumer surveys. Customers surface as individuals, however, as well as in the form of statistics. They meet and talk with the manufacturers' consultants and the retailers' sales staff and so give these representatives of the big actors an idea of their likes and needs. They call in on

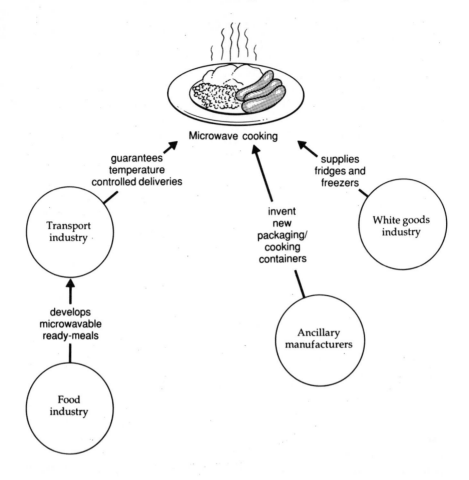

Figure 1.7 *Industries generating new supportive products or services are important actors in the innovating world.*

Customer Relations phone lines for advice or to register a complaint. As users they are in touch sometimes with after-sales service engineers, who also pick up from them the nature of problems with microwaving in practice and feed it back to retailers and manufacturers.

The purchase of a microwave is only a short-run token of the public's effective enrolment in the microwave cooking project. They must also, to secure the future of microwaving, vote with their cooking and eating habits. Such changes hang less on any one manufacturer's artifacts than on two other kinds of development. The first is changes in social relations in society generally – such as more women going out to work and more independence among teenage offspring. These social concomitants of microwave cooking will be discussed in Chapter 5. The second kind of development needed is the presence, the provision, of a very particular infrastructure, sketched in Figure 1.7.

In the UK it goes without saying that a widespread national electricity supply is a pre-requisite. In some countries the time is not ripe for the microwave project if for no other reason than that many households do not yet have electricity with a stable current and affordable charges. In addition, though, microwaving prospers in partnership with other technologies (see Figure 1.7). The plastics and packaging industries must be generating microwavable containers (trays, sachets), and appropriate wrappings for microwave foods. Enough households must have freezers. Although a microwave oven can be used in the absence of a freezer it is more likely to be wanted and used in a household that has one. It helps if the food industry is developing branches that produce not only interesting frozen ready-meals, but also high-quality cook-chill recipe dishes (short life) and, another recent innovation in food technology, ambient temperature food options (long life). These are, as we shall see, in natural alliance with the microwave oven project, and some have actively promoted microwave cooking through membership of the Microwave Association, running advisory services and sponsoring awards.

For the cook-chill innovations to take off, highly specialized and efficient temperature-controlled transport and storage systems must first be developed. The recipe-dishes manager of a leading food retailer explained how her products would be in course of preparation on a Wednesday morning; by Thursday morning they would be cooked, chilled and packed. This speed of preparation is little use, however, unless by Thursday night at 10 p.m. these dishes have travelled, in prime condition, to a regional depot so that by 6 a.m. on Friday morning they can be in the shops ready for the customer who calls before work at 8.30.

In the UK during the 1980s these conditions materialized. The total market for reheatable convenience foods in the UK, France, West Germany, the Netherlands and Belgium grew by 41 per cent between 1985 and 1988, and appeared set to continue climbing. Just as the microwave oven boom took off first in the UK, so did the habit of buying reheatable convenience foods. The United Kingdom dominated European sales in the late eighties, with a figure almost as high as that of the four other major buying nations combined. The UK is also characterized by the high percentage of such sales that are for household as opposed to catering use (Marketpower Ltd, 1989). Particularly popular in the UK are quality cook-chill dishes, with many of the major food supermarket chains competing by offering their own brands. Though most ready-meals may, at some disadvantage in time and energy expended, be heated in a conventional oven, some food manufacturers are now bringing out lines of ready-meals that are specifically labelled as for microwaving only.

All of these developments, infrastructural in relation to the oven manufacturers' project, have moved forward fast in the late 1980s and early 1990s, producing an optimistic scenario for the ultimate goal: an extensive practice of microwave *use*. It may not be precisely the use manufacturers have in mind. The interpretive flexibility of the microwave oven may persist to the point of consumption: we must retain an open mind on this till Chapter 5. The provision of certain features does not guarantee they will be employed.

The foregoing account shows that a technological innovation such as microwave cooking is not the brain-child of some genius. It is the creation of a complicated network of actors, individuals and organizations – and at the same time the project itself creates the network. This represents one version of the story of the forces producing a new cooking technology. We have chosen to use concepts from theories of the social construction of technology to provide a focus, an arena and a language. It is only one of many alternative, equally valid, stories that can be told from different standpoints about microwaving: the engineer's story, the economist's story. Ours is first and foremost a *sociologist's* account. It looks at the *social relations* that generate, impel or impede the innovation.

As sociology, however, is the story complete? Bruno Latour says that 'understanding *what* facts and machines are is the same task as understanding *who* the people are' (Latour, 1987: 140). We have a picture in this chapter of many of the people. Yet the preceding account, as is quite normal in social studies of technology, avoids any reference to subjectivity and identity, and it fails to gender any of the actors: no 'he' or 'she' has yet appeared.

Much more can be learned by asking whether the actors are women or men, whether phenomena are represented as feminine or masculine. Once we think about gender we are obliged to think about people and groups in terms of *how they see themselves and each other*. And it will become clear that gender subjectivity and identity have an important bearing on technologies and technological activities. People as feminine and masculine are also constituted in relations, some of which are technological relations.

2

Gender in the Microwave-World

In Chapter 1, the microwave-world was described without any reference to the sex of the actors. This is normal enough in social studies of technology. It is not felt to matter whether the sexes play different parts, or, if so, which sex plays what part. A feminist analysis on the other hand suggests that a person's sex almost always counts. It makes a social difference to any situation. We are suggesting in this book that different aspects of the social shaping and social shape of technology become apparent if we have eyes for gender and that, besides, we shall learn something more about gender by seeing it through the prism of technology. Our next step therefore must be to develop our metaphor of the microwave actor-world, detailing it as a *landscape* with areas and gradients. We need to colour in a contoured map of actors and show where women and men are to be found, the gender pattern of location.

We are concerned at this point first and foremost with numbers and activities: who does what. From this picture we shall be able to understand the *topography* of the separation of the sexes (the categories of activities and positions along which the fault lines run), the *rationale* behind it, how people *represent* the reasons to each other. We shall begin to get a sense of the relationship between material differentiation of the sexes and symbolic differentiation. Secondly, we shall see what practical inequalities between women and men are associated with the technological sexual division of labour.

As we have seen, two major organizations, one producing, one distributing, lie at the heart of the microwave-world. We shall look in some detail at the gender pattern of work in each of these and through them show something too of the place of the sexes in the organizations of the wider world with which they are in contact. We begin with Electro Corporation, looking first at the very coalface in this microwave story, the production operation, before going on to consider the broader functions of product planning and engineering.

Sexual Divisions in Microwave Production

Electro UK's microwave ovens are engineered and manufactured in the corporation's rural factory, and this is as good a place as any to begin the gendering of the microwave-world. Seen from the nearby range of wooded hills the plant looks impossibly large. White, foursquare and featureless it resembles nothing so much as a flattened microwave oven. Local people do not appreciate its utterly modern style. Annoyed by the perimeter floodlights that by night illuminate the site like a high-security prison, they say 'the coal pit and its slag heaps were a black eyesore. We've just exchanged that for a white eyesore.'

But the jobs the Japanese have brought are badly needed. The local labour force has been used to finding its livelihood either in farming or in the declining industries of the area, coal and steel for men, sewing, packing and service work for women. The recent deals with incoming Japanese electronics firms have provided a new opportunity for employment for both sexes. The local council were afraid the Japanese might offer only low-paid, unskilled work suitable for women. Brokering a deal with Electro Corporation that included a single union match with the heavily masculine EETPU, they called on the firm to recruit a 'balanced' workforce.

They need not have worried: Electro recruitment at its new site has been, in overall numbers, favourable to men. At the time of our visits the breakdown of a total workforce of 1124 was as shown in Table 2.1.

Table 2.1 *Electro UK plant workforce**

	Men %	Women %	Occpns %
Direct workers	60	40	76
Forepersons	67	33	7
Supervisors	86	14	5
Technicians	97	3	6
Apprentices	89	11	2
Assistant managers	85	15	2
Managers	91	9	2
Total workforce	66	34	100

* In this and ensuing tables the first and second columns show the percentage men and women comprise of the employees in each occupation; the third column shows the percentage each occupation represents of the total workforce under consideration.

The fault line here is clear: controlling jobs and technical jobs are more masculine than subordinate positions and jobs not involving technological know-how. Referring to the microwave section, as one woman put it, 'all the technicians and engineers are male . . . everybody who is at any level is male in microwave'. The company does not have a conscious sex equality policy. The assistant personnel manager (a woman) said, 'We don't have anything written down. It's just basically, we just try and get the best people for the company.' Photographs running through this chapter show something of the relation of employees to each other and their technological jobs in the design and manufacture of microwave ovens.

Of the direct workers 130 are employed on microwave production, while the others are manufacturing business equipment. The sex-ratio on the microwave line has gradually shifted over time. The company began with a preference (as in Japan) for a majority on the line of young women of 16–19 years. When the plant opened in 1986 the proportion was 70F:30M. Recently, however, men had inched up to around 50 per cent and the average age of the remaining women had increased too. The company was now advertising specifically for women up to 50 years of age.

The reasons given for having more men and older women on microwave production differ in kind. It is said that the closure of the local steel plant and other sources of male unemployment had led to more men applying for these 'feminine' jobs: the ratio among job applicants was currently 70M:30F. As to the age of women, the very rapid turnover of 'girls' had been a source of annoyance to management. The culture of discipline and loyalty, both in the firm and the family, that often holds Japanese employees at their workplace for a lifetime, does not pertain in Britain. As the production manager put it, 'If you're 16 and you don't like it today because the supervisor shouted at you, you'll go home and Mum will say "tell 'em you're not going to work".' At one period when the demand for microwave ovens called for extra production the company departed from its normal practice and took on older women as temporary part-timers. Finding them surprisingly good, the Japanese management decided to adopt British habits in future. They continued, however, to employ everybody full-time and to avoid casual labour.

To the visitor the plant gives an impression of order and discipline. Motivating messages are prominently displayed. One reads: 'Quality First In Heart And Mind. Stronger Commitment To People. Greater Dedication To Product Quality And Value.' The Japanese philosophy is that product quality and vigorous effort make for all-round spiritual and material wealth, benefiting employees and customers alike. Scanning the large, well-lit space, accustoming oneself to the noise, you begin to notice the people. Here in an aisle a uniformed Japanese manager is in discussion with a British subordinate. There behind a glass screen is another, preoccupied with paper work. The floor is filled with operatives at their work stations, some of which are in the mainstream of the conveyor system, others in little by-waters of sub-assembly.

The first part to enter the construction process is the shiny metal 'cavity' that is the structural frame of the oven. As it moves from 'cavity feed' through 'main assembly' the cavity gains a magnetron, motor, fan, turntable. In 'electrical' many bundles of wires and other little components are inserted and attached. The control panel and door are added. The enamelled outer wrap goes on. Now it really begins to look like the microwave oven we know. In 'finishing' it is tested, polished, instruction labels are stuck on and manuals put inside. The oven moves onward to packing, where it is protected with polystyrene and stuffed into pre-printed cardboard boxes before being stored in the warehouse.

The balancing of the production line is a fine art, the responsibility of the Production Engineering department. Assembly must be organized into activities of matching length. While we were in the plant the task time was 29 seconds. An optimum, managers believe, is 2.5 minutes' duration. Any less and the workers become demoralized; any more and concentration is lost, inefficiency results. The line stands still while the tasks are completed. With five seconds to go a warning 'music' sounds over the loudspeaker – an annoying little sequence of notes that, experienced all day long, addles the mind as much as any other aspect of these punishing labour processes. The worker hastens to complete his or her task. The line moves on, and yet another partially assembled oven replaces the one that went before. Each oven was taking 35 minutes to come into being. Up to a thousand were coming off the line each working day.

The line runs for two and a quarter hours without a break, the only intervals being a 30-minute lunchbreak and ten minutes for tea. Anyone wishing to leave the line to go to the toilet must call for a substitute. A 43-hour week chained to this line generates strain of many kinds: of back, shoulders, arms, wrists. Above all it takes a heavy toll on the mind and spirit. It is not surprising that employee turnover here is more than 35 per cent per annum. People do not stay longer than they have to.

Looking up and down the line one forms certain first impressions of the gender topography. Both sexes wear a similar blue overall and are visible in not unequal numbers, but it is important to remember that this *is* essentially women's place in the company: nine out of ten women working at this plant are here. The next impression is that the most likely place to see a woman is next to another woman, the most likely place to see a man is next to another man. Men are stretching and reaching, lifting the cavities onto the line, installing the heavy transformers, taking wraps off the overhead conveyor. They are doing the jobs for which you need to stand, and those for which heavier tools are required. Only men are to be seen in the packing area, the warehouse and operating the printing presses. Of the people moving about, most are men. Men stand behind women on the line, discussing and solving technical problems. Women mainly sit. They use their hands alone or small tools – pliers, screwdrivers. They twist their fingers into small spaces to find elusive fixings and bend their heads to see inside the cavity. Yet here and there are individuals who interrupt the pattern: a young man sits with the women in sub-assembly, a woman supervisor moves up and down the line.

What is the *rationale* here? Managers, and the women and men themselves, explained the pattern as resulting from sex differences in people's qualities, and skills matched to demands of the various tasks. Dexterity was undoubtedly women's primary attribute. Women were described as suitable for 'fiddly' jobs, where you had to 'get your fingers in there'. They were seen as relatively tolerant of boredom and routine, and therefore suitable for immobile and repetitive tasks. Women, by virtue of their family role, were represented as particularly careful: 'I think, whether it's maternal instincts with some of them, I don't know, whether they *mother* the things or anything, I don't know. I think they're more conscientious . . .'

Women, said this male informant, had more 'will' to do menial tasks well. They were more biddable, and therefore easier to manage. The production manager said of operators:

> males will tend to put their own interpretation on what is required, where a woman will do what she's asked to do. So, for the type of industry, a woman is far more suitable for the light assembly side than is a male.

As we saw in Table 2.1, women are not excluded from first-line supervision. This too is because they have certain social skills men are seen as lacking. 'That's where females come into their own, I suppose,' said an assistant manager in Production Engineering. 'They take on this new operator and make them feel at home. Whereas you've got your – your, how can I say it, your more brusque and blunt lads, "right mate, this is what you do".'

Men are represented as taller and stronger than women. This determines their place in heavier work. 'Moving microwave boxes about, you would, oh, if you've got a particularly big strapping girl, fine. But you wouldn't *normally* look for a woman in that job.' As another man put it, you'd not want the girls to 'come out with muscles like Arnold Schwarzenegger and saying that they can't tap off with the boys'. Men are seen as liable to seek greater mental stimulation in their jobs, as being mentally tougher and carrying greater authority. 'Women don't like working for other women,' said a manager in production. 'Women often prefer to work for men. Men probably prefer to work for men . . . It's this perception that, you know, excuse the French, the women haven't got the *balls* for the job.' Above all, men are seen as having a bent for technology that women lack. One manager said:

> I do find that, why I don't know, but women tend to be a little bit frightened of working with electricity . . . I think it's to do with, going back to school and what have you. Women – well, it's only very progressive schools where girls are doing metal work or what have you.

This issue of the gendering of technical competence is discussed at greater length in Chapter 3.

The occupational sex-segregation was also represented by management as not only arising by management policy but being in part elective: 'I think the men, it's probably not all down to stature. I think the men like to be where all the other blokes are.' Women, for their part, are happiest working together where they 'can have a bit of a chinwag'. It was also interesting to see how these widely held preconceptions were not entirely inflexible. Management seemed willing within reason to allow individuals to try non-stereotyped tasks. So two young women had asked to go on cavity feed, where they were found to be quite satisfactory. A man had settled himself into the fiddly job of fixing door hinge brackets, a job he seemed to prefer, and was consequently described as having the quality of 'nimbleness' normally ascribed to women.

The overriding criterion was not gender-tidiness but what worked best. Management had, by all possible means, to resolve the basic contradiction of this kind of production. As one British manager put it, 'What the Japanese have done is, to sum it up, they've deskilled the work, by breaking it down into smaller and smaller parts.' Such Fordist techniques call for an all but impossible combination of thoughtful responsibility with mindless oblivion. 'You need some thickies basically, but you need them intelligent enough to understand that it's important.' Management were therefore prone to be pragmatic. If 'you're getting a better standard of woman, for less money, than of men', as managers observed, this disposed towards the recruitment of women. If men now presented themselves for jobs in greater numbers, that tilted recruitment the other way.

The women we interviewed, however, were clear that when men turned up beside them in 'women's work' on the line, you did not expect this to last. 'More often than not they're here until they find something better . . . where they can use a bit of their, erm, brain. Because a lot of them feel it's brainless. You don't need a brain to work here.' Women were quite certain they faced prejudice in this company, that women were not considered suitable material for training and advancement to production management. 'It's definitely the males, the males who get the promotion.' 'They're all men. It's very male oriented . . . There's a machine – a man'll run it . . . It's the men, men, men.' 'They tend to promote men more because they think, the Japanese especially, think "Oh, she's going to have another baby".' Some do put the discrimination down to the Japanese – for by reputation the workforce at the corporation's headquarters in Japan was profoundly sex-stereotyped. Others said that British managers had their full quotient of sexism. One male manager, Tom, supported the women's view, and criticized in our presence the (British) production manager's attitude to women.

What was at issue in this disagreement was this manager's failure to promote a particular woman, Carol, to be the microwave department's first woman technical supervisor. Carol may well stand as symptomatic of the difficulties women face in Electro in obtaining promotion off the line, especially in a technical direction. (By 'symptomatic' as used here and elsewhere we mean that this woman's experience is not discrepant with that of other women in Electro's production area, while, were she a man, such experience would be discrepant with that of other men.) She was a mature woman who had joined the firm along with others like herself on a twilight shift to boost microwave production at a moment of peak demand. She had for years been a sewing machinist on piecework in the local garment trade. There, however hard she had pushed herself, a progressive deterioration in rates of pay had meant she

could never earn more than £100 a week. The factory had closed down, not because it was unprofitable, but because of 'labour problems': its workers not surprisingly deserted it when companies like Electro offered an alternative. Here the pay, though not generous, was a little better. She could earn over £120 a week with bonuses and the work was not as hard on her shoulders.

Kept on as a full-timer she was put in the 'finishing section' doing work she described as cosmetic, tedious and undemanding. Then she was moved to the electrical section. 'Which was a new challenge. There was – at least you know more of how the oven worked. I mean, before I went on there, power consumption was something, well it was just a word. Well, I fully enjoyed it on electrical.' Four months later she was made foreperson and then, a step further, acting supervisor. In charge of 16 people, Carol's responsibilities included motivating them, ensuring correct line balancing, seeing that the best methods of work were used, and monitoring the quality of output. She was put on an internal training course of the Institute of Supervisory Management.

But when the time came for a permanent appointment to be made to the supervisory job in which she was currently acting, Carol was passed over for a younger man. She was bitterly disappointed. The production manager explained his reasons to us. 'It was a technical role looking after equipment, getting under machines and fixing machines, so we told her and she understood, you know, that she wasn't the type of individual for that. She's better off controlling people.' Tom, the production engineering manager who had been critical of this decision, said he felt insufficient effort was being made to identify competent women and get them off the line. The reason: management believed 'it's machines that are too technical for women'. This was borne out by several others who remarked that women on the line were more likely to be promoted as 'junior leader' than as technician.

One of these was an operator called Don. He stands as a useful contrast to Carol, equally symptomatic of men on the microwave line as she is of women. He came straight from school, intending it as a stop-gap before the Army. Eighteen years old when we met him, he worked in electrical testing and had just been sent by Electro to a local technical college on day release to take his City & Guilds exam in electronics and electrical studies. Dissatisfied with the work, he was ambitious and banked on being noticed. 'I mean, you can come in here without any qualifications whatsoever and still make it pretty big.' Indeed several young men spoke of aspirations to senior management in confident tones we heard no women use.

We can see here something of the way differentiated gender identities formed in the wider world are reinforced and exploited in the technological relations of the microwave-world. In turn, meanings made of these experiences will travel out and have their effects beyond the factory gates.

Designing the Product: Male and Female Roles

Stepping away from the line now, in Production Engineering we did find two local women, both in their early twenties, who had arrived as operators, had followed the traditional women's route to 'junior leader' but had then found themselves on a less usual track. Wendy had been regraded as 'junior technician' and Karen appointed 'work study technician'. These, with the exception of a sponsored engineering student here for a few weeks in her holidays, were the most 'technically' skilled women we found associated with microwave activities in Electro. Karen's job involved going out on the line and doing time and motion studies of people at work. At times, when 'activity sampling', she had to spend a whole day timing one person, including their trips to the toilet – something she found invidious. Wendy worked in the same team as Karen, breaking down the production process into balanced tasks and writing instruction cards to be posted above each work station.

What was interesting to us was the way both young women despite being called 'technicians' described their work as *non*-technical. Karen said, '*technical* is something like manual work . . . What I'd call technical is *repairing* the microwaves, something like that.' At first they had expected her to do the typing and tea-making for the production engineering department. She had resisted this, but still her own job did not seem like the work she saw her male colleagues doing. Wendy said of her own job that if it had any connection with technology it was not with the microwave and its production but with the office computer. She felt she would be better termed 'information analyst'. The work called less for technical knowledge than 'really just common-sense'. There were gaps in her knowledge, she felt, about how the microwave 'really works'.

> Well, we've got somebody with electrical knowledge anyway in our department. So I take a back-seat, then, and he does any of the electrical working instructions . . . that's his, that's his speciality. Mine's just mainly the build, which I feel anybody can do.

We felt that the reluctance of Wendy and Karen to claim technologist status partly reflected an observable fact: their jobs had more to do with people and less to do with things than those of the male technicians. Yet they seemed also to be expressing a characteristically feminine doubt that anything *they* could do should be graced with the name engineering. What was at stake were the boundaries of 'technology' and femininity. In a way they were reiterating the stereotype of the incompatibility of these two terms.

Apart from Wendy and Karen, the non-production departments of Electro UK were devoid of women engineers and technicians. The Japanese manager heading both design and production engineering explained that his engineers had to be all-rounders and whereas 'I feel women give very nice performance for printed matter, semi-conductor

design work, very deep and careful, detailed work, but the microwaves are very *heavy*'. The result was that he headed a design team of bright young British males. One of them said he saw no reason why a woman wouldn't fit. However, since 'they just don't seem to apply for the work', this open-mindedness was never put to the test.

Of course this engineer was certainly correct in observing that in the wider society there are few women engineers from whom to recruit. In those areas of the actor-world adjacent to Electro Corporation's engineering functions – the parts vendors, the microwave specialists, colleagues in rival firms – all the qualified personnel (they reported) were male. One specialist we interviewed cast his mind over the world-wide microwave specialist engineering community to which he belongs – around 70–80 people, at his reckoning – and could think of no woman among them. On the industrial design side, it is known that in the UK 99 of every 100 industrial designers employed in industry are males (Sale, 1988). Within Electro, besides, engineering coloured many associated functions and made them masculine too. Even in the marketing and planning section, apart from a female secretary, the staff were male. Its manager was about to recruit a new team member and mused that it would be 'nice' if it could be a woman. But engineering was a necessary qualification. 'Because the interface is so close with the factory, somebody without a technical background would have difficulty in having a relationship with the various departments that really provide the most essential and critical work through product design.' The result of this association of the microwave with technology, and technology with masculinity, is that in Electro Corporation's non-production departments women were few. Through similar processes the computer department, described by one woman as 'home of the whizz-kids of Electro', was also a man's world, though neither microwave engineering nor heavy lifting was involved.

Masculine too was the upper management structure. Men alone occupied the four top grades. Women featured only as assistant managers in such functions as personnel and payroll. We quickly learned that it is not gender alone but ethnicity too that gives feature to the topography of Electro Corporation. As one woman put it, looking up the hierarchy, it is not just men and women you must distinguish, 'it's women, men and Japanese'. The Japanese in question were, however, all male. The managing director of Electro UK was of course Japanese. So too, immediately beneath him, were the director of the factory and the director of Product Planning and Sales. Only at the next level down did British males enter the picture as, for instance, production manager, manager of sales administration and the product manager for micro-waves. The principle was that Japanese should maintain control of overall direction and policy-making.

A second principle at Electro UK was, as the British managers under-stood it, that Japanese personnel should control engineering knowledge.

Thus the deputy general manager, also in charge of Quality Control, and the director of Design Engineering were Japanese, while posts such as purchasing manager and finance manager, as well as responsibility for such activities as parts control and inspection, could safely be left to the locals. All important engineering decisions originated in Japan, with information and instructions arriving by fax and telex in the Japanese language, which the British could not read. Although, as we saw in Chapter 1, Design Engineering in the UK is now being given a more active role in origination, the British engineers and managers continued to speak bitterly of a Japanese monopoly of technological know-how, with diffusion of technical knowledge limited by the Japanese to 'tell me what you want to know'. British managers felt their strongest card with the Japanese was their cultural affinity with the British workforce which made them the most appropriate people-handlers. Masculinities were being shaped here, around authority and technology, in distinct ethnic forms.

British women, it at first appeared, had no strong card to play and filled the roles they commonly fill in organizations in both Japan and Britain: the clerical workforce, the office supervisors and lower management. There was, however, one interesting exception. This had to do with activities where *cooking* knowledge was needed. We uncovered two groups of women with such cooking knowledge: the home economists of the Test Kitchen, associated with Product Planning; and a team of 'consultants', or demonstrators, assisting the sales effort.

The home economists numbered three, and all were full-timers. Two worked at head office, one at the plant. Just as engineers are recruited from a national pool that is largely male, home economists are drawn from a mainly female population. Our research in the actor-world of the microwave took us to the Microwave Association, to home economics departments in further education colleges, to the editorial offices of household magazines and into the world of the microwave cooking writers who invent new recipes and publish books on the new cookery. Few indeed of the home economists are men. It is therefore not surprising that those of the profession who are recruited by microwave manufacturers are women too. Our Electro home economists reported that all of the colleagues from competitor firms they met on the microwave panel of the manufacturers' association were women.

As to the microwave consultants, they too were women: 100 part-time employees, coordinated and led by ten coordinators and four full-time women 'microwave specialists'. The sales team, on the other hand, comprised 14 men and one woman. The manager of both groups, a British man, saw the consultants as 'a sideline'. A home economist told us she felt the consultants' skills could and should be acknowledged and used as a selling resource. Their manager's explanation of the separation into two teams with two functions confused economic rationality (you had to pay sales staff more) with the need to reflect within the company the sexual division of roles in the world outside:

We decided that salesmen are salesmen because that's what we're paying them to do. We don't want the housewife badgered by salesmen. We want the housewife to feel comfortable and confident . . . So by keeping them totally separate and devoid from each other, we maintain that very important role.

In cooking roles in Electro, then, no men were to be seen. On the other hand, the managers to whom the cooking women were responsible were men. Gender difference here was being clearly constituted round contrasting abilities ascribed unequal value. The highly value-laden distinction between cooking knowledge and engineering knowledge is explored in greater depth in Chapter 3.

Our distance from Japan made it impossible for us to acquire precise information on the sexual division of labour in Electro Japan. Accounts we received both from Japanese managers in Britain and British managers who had visited Japan suggested that the occupational sex-segregation there, both horizontal (type of job) and vertical (level of job), was yet more complete than in Electro UK. In particular, a British manager observed that engineering work was '99.9 per cent male . . . but to my knowledge there's no male home economists'. As to management positions a woman said, 'They're not particularly keen I think on having the ladies tell the men what to do. It doesn't go with the whole of their history.' A further factor exacerbated women's disadvantage in the Japanese context: a reluctance to employ women after marriage. We were told for instance that the large Test Kitchen in Japan was staffed mainly by women who were unmarried. Although women in Japan, like women in the West, are beginning to make demands on the system and a few now achieve careers comparable to those of men, the corporate climate is not welcoming. In particular the expectation that local employees will work exaggeratedly long hours with negligible holidays makes such employment all but impossible for women with domestic responsibilities.

Electro UK is a hybrid of Japanese and British organizational culture. The Japanese attempted to import their preferred style, involving total commitment, quiet deference, consensual decision-making. Those managerial and professional Britishers who had to work with the Japanese strangers, not only in Electro UK but in other organizations with which they had dealings, did in some cases tentatively adapt their behaviour to suit what they felt to be the strangers' expectations. Thus Electro UK's advertising agency had decided to appoint male account executives to deal with Electro's Japanese managers, fearing that a woman would not be welcome to them.

The Japanese found the response of the manual workers particularly disappointing. They had begun by instituting special shoes for work, callisthenics for keeping fit, quality circles for involving workers in product excellence. They expected loyalty and pride in belonging. They soon acknowledged however that Japanese culture could not be imposed. 'Electro is my *life*,' said one. '[But] I don't want to – I don't think I

can change their minds. No.' So they adapted their management strategy, with the help of British managers, to something closer to British expectations. The trim egalitarianism of the blue uniforms remains. The practice of marginalizing the trade union and relying on the company-controlled staff association for industrial relations reflects Japanese preference. Given their tolerance of local idiosyncrasy, the Japanese might well have accepted a strong role and presence for women in management and engineering jobs in Electro UK, had the British organizational culture suggested this. The presence of Wendy and Karen, and the appointment of the female sponsored student, though British managers had felt they must get clearance for such innovations from their Japanese superiors, had not been contested by them. British management in Electro was no more egalitarian than most British enterprises. When forced to a conscious position managers professed to endorse 'equality', but in practice they actively reinforced gender stereotyping.

The Gender Topography of Retailing

A mapping like this of the presence and absence of women and men in the various organizations, departments and roles concerned in the origination of microwave ovens reveals a distinctive topography in which the

peaks and pinnacles are mostly male while the middle reaches vary according to the nature of their business. Women are found higher up the slopes when the hill in question is Product Planning, with its Test Kitchen, but they hardly creep above sea level in Computer Department. On the other hand, the lowlands are patchy, with men clustered in some departments and activities, women in others. The line that separates senior from junior, authority from subordination, like a tree line or a snow line on the mountainside, is a clear divider of the sexes. Other fault lines exist that demarcate groups of a similar level but different function: concern with things v. concern with people, engineering v. cooking, project v. routine. These are some of the things that seem to determine whether a job will be done by a man or a woman. Technology is a clear orienting factor in location and allocation. Let us turn to Home-Tec and see whether its gender topography follows similar principles.

A firm such as Home-Tec is logically divided into three major sections: head office functions, distribution and retailing. We can consider each in turn. Several hundred people were working in head office at the time of our study, and the share of women and men in this total was not far from equal (see Table 2.2).

However, almost one-third of the head office women employees were clerical and secretarial workers, while almost half of the men were in middle or senior management, a skewed distribution of the sexes not

Table 2.2   *Home-Tec's head office personnel*

	Men %	Women %	Occpns %
Managers:			
Senior	92	8	7
Middle	83	17	22
Junior	61	39	39
Clerical/secretarial	14	86	32
Total workforce	53	47	100

uncharacteristic of British employment. 'It's male biased,' said a woman manager. 'The top management are all men . . . you have to be exceptional, more exceptional than a man. It's more difficult for a woman to get on than a man.' And she added in a sarcastic tone, 'There's always the fear she'll go off and have babies I suppose.' A male manager confirmed, 'This company, I think . . . could be accused of being the type of company that gives management roles to men more than women. Certainly the higher you go up in the company there are less and less women.' Photographs in Chapter 4 show something of the gendered roles and relationships involved in this kind of retailing enterprise.

Head office was divided into several major divisions. Some differences in the presence and position of women were apparent between them, depending on their function. Finance, for instance, employed very large numbers of women in junior management and clerical grades. The commercial division contained the vital functions of marketing and buying. Both had their fair share of women, but they mainly dealt 'with what you'd expect women to know about . . . the white goods'. Yet the microwave buyer, the person most in contact with Electro UK and other microwave manufacturers, was a man. Computing, here as in Electro, was heavily male.

> If you go down to our management information services department – very few women there . . . The people who know most about the system are men. The people who [*pause*] *don't* know very much about the system other than how to put something on to it, or how to find something, are the women.

The Customer Services Department, by contrast, employing four times as many women as men, had a female manager. The business planning section had been staffed by three women, including its manager, and one man. One of the women had left, and a preference was now being expressed for the vacancy to be filled by a man in order 'to dilute the atmosphere'. It is significant that in those few areas of management where women predominated, a corrective rebalancing of the sexes was felt to be needed. Nobody expressed a desire to strike a better balance where men were numerically dominant. Nor was balance a major issue where women were numerous at the lower levels.

The staffing of the distribution centres and warehouses is shown in Table 2.3. Overall, men predominated across the network, comprising 74 per cent of the workforce. No women were employed at all in the higher management grades and they were heavily outnumbered even in junior management.

Table 2.3   *Home-Tec's distribution workforce*

	Men %	Women %	Occpns %
Managers:			
Upper	100	0	1
Middle	83	17	3
Lower	71	29	17
Administration	15	85	21
Warehouse	89	11	26
Maintenance	100	0	1
Drivers	100	0	31
Total workforce	74	26	100

Drivers were almost one-third of the distribution workforce, and none were women. Eleven per cent of the warehouse staff were female. The figures did not specify their role. However, our parallel study of a similar company, Wonderworld, showed warehouse women to be in clerical roles while the men were goods-handlers.

Shoppers out to buy a microwave oven see nothing of the Home-Tec staff so far described. Home-Tec to them begins and ends with the shop and the sales assistant. It is the staff of company's hundreds of *stores* that are the third major division of personnel and, with a staff of several thousand, they account for three times as many employees as head office and distribution together. Table 2.4 shows the sex breakdown for full-timers, Table 2.5 that for part-timers, who represent one-third of all retail personnel.

Table 2.4   *Home-Tec's full-time retail personnel*

	Men %	Women %	Occpns %
Stores (Arrow)	82	18	47
Stores (Bunnett's):			
high street	79	21	36
superstores	79	21	17
Total workforce	81	19	100

Men comprise the great majority of the full-time retail workforce. The company runs two distinct chains. The stores we call Arrow were the original Home-Tec stores, specializing in leisure electronics. Here the

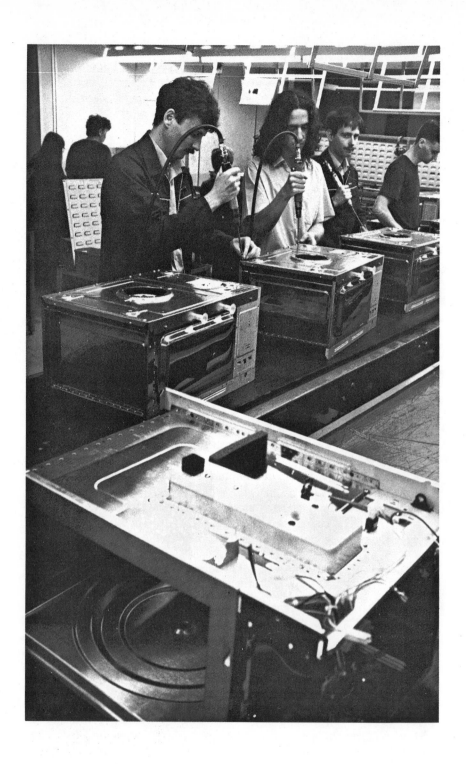

majority of men was even more pronounced than in the Bunnett's chain, acquired in the take-over, where kitchen and other functional domestic equipment was also sold. Within the figures we found a distinct age pattern: in all store types the men tended to be younger, the women older. This applied even more markedly among the part-time staff.

Table 2.5 *Home-Tec's part-time retail staff*

	Men %	Women %	Age groups %
Under 19 yrs	72	28	13
19–25 yrs	50	50	18
26–39 yrs	8	92	36
Over 40 yrs	11	89	33
Total workforce	25	75	100

The young male part-timers were, we were told, mainly Saturday staff, who were likely to have some other full-time occupation, possibly studying. Though there were also Saturday women, the great bulk of women part-timers were working a short day or a short week.

We were not permitted to obtain figures from Home-Tec showing the relative numbers of women and men who were divisional, regional and branch managers. We were told anecdotally that there were no women divisional managers, possibly one out of approximately one hundred regional general managers and 'only a few' women branch managers. Our information from Wonderworld, Home-Tec's main competitor among the multiples, confirmed the predominance of men: they formed 95 per cent of all store managers.

Once in the shops, the microwave oven takes its place among the other housework appliances as a 'white good'. Our case study was of a branch that was in most respects a typical Bunnett's store, selling both brown goods and white goods. Situated in the centre of a middle-sized town, it was nonetheless classified as a superstore – making it something of a hybrid. In staffing too it was characteristic. The manager was male. Of his thirteen staff, eight were men, all full-timers. Of the five women staff, four were part-timers (in line with the company target of a proportion of seven full-timers to three part-timers). Also characteristic was the deployment of the sexes in relation to the jobs within the store. That the branch manager was male surprised nobody. His regional manager described the profile required for this job not only using the masculine pronoun but in strikingly masculine terms. The question you ask yourself is:

> can he motivate, can he train, can he lead, can he handle five hundred things at once, has he got a flair for merchandising, can he understand stock control, and can he deal with customers you know, etcetera, etcetera? I mean it's a

helluva thing. It's a helluva difficult role and the trouble is it changes.

The store-keeper too was a man. His job involved continuous 'man-handling' of goods inward and outward, using a forklift and other equipment. 'It's really not the kind of job your average woman would want to do,' said the manager.

> I'm not saying she couldn't do it. Because we saw only the other day a lady [from a vendor firm] delivering washing machines on a big articulated lorry and, er, we were amazed. And she certainly didn't need any help . . . [but] the stores job is ideally suited to a man.

The labour process on the shop floor could hardly be described as punishing in the manner that Electro's microwave assembly line is punishing. Nonetheless it had its own peculiar pressures. Managers were very much aware that 'people are your best asset': every person on the shop floor added to turnover, and a good person added disproportionately. Getting good people for minimum pay, motivating them by incentives and equipping them by training, these were the store manager's daily project. Every sales assistant was assigned a target: she or he must aim for so many thousand pounds' turnover per week. A low wage could thus be supplemented by a monthly bonus: £4 or more per £1000-worth of sales, depending on the total attained. The sales staff were offered special deals to increase the incentive – certain products at certain times would earn a higher bonus than others. Assistants were expected to take only 15–20 minutes' lunchbreak and most of them, thinking of their pay packet, complied. The Electronic Point-of-Sale computer system which links the till in each store to a mainframe computer at head office not only organizes stocks and supplies. On registering a sale it also works out the commission due to the salesperson concerned and transfers the information to the payroll database.

To boost sales, the staff were subject to continual training towards CORE – the Certificate of Retail Excellence. It involved both sales techniques and product knowledge, categorized as Features, Advantage and Benefits. Catch-phrases assisted the inculcation: 'FAB', 'It's up to me. It's got to be me.' Achievements were reviewed weekly and 'you get an ass-kicking if it isn't right, basically'.

Though selling was meant to be unisex, in practice men tended to gravitate to the sale of brown goods and women to the sale of white goods.

> No matter what we do to try and avoid it [said a training manager], it just naturally seems to happen that in [Bunnett's] stores the women will stay with white goods and the men will stay with brown goods. You do get interaction and cross-over of course, but that just seems to be the way it goes. It's very unlikely that you'll get a sixteen, seventeen-year-old boy who will go on to washing machines. Its sort of familiarity. He's probably never used his mum's washing machine but he has got a hi-fi.

Conversely, women are seen as, and see themselves as, 'having a feel for'

cookers, microwaves and other white goods. 'It's still this stereotyped image if you like, that white goods is female and brown goods is male.'

We shall see in Chapter 4 that head office is trying to break down this stereotyped image and practice. For the moment, however, it prevailed in our case study store, and indeed there was seen to be some commercial advantage in it. The typical customer in an Arrow (brown goods only) store and on the brown goods side of a Bunnett's store was visualized as a young man. It was assumed that he would expect the best sales advice (in response to what were anticipated as technical questions) to come from a young man similar to himself. Shoppers for white goods were assumed to be women or couples. Although male technical authority was presumed to span both product ranges, there was a feeling that when it came to *use* of a cooker or a microwave, the woman would value the domestic wisdom of a woman sales assistant. The gendering of brown and white goods of course is not peculiar to Home-Tec. It is discussed more fully in Chapter 4.

Two other daily tasks shared among the salespeople also revealed a sex difference. Staff were responsible for cleaning and tidying a section of

the displays. In practice some of the men did not always clean their areas and a female would clean for them. During our observation in the store we saw women but not men carrying out this routine dusting, although one elderly male salesperson was observed polishing washing machines prior to displaying them in the showroom. A second task was setting up and tuning in TVs and VCRs for display, a job done when there were few customers in the shop. No female sales staff were observed doing this. Some, we were told, had on occasion done so, and it had been considered a significant achievement.

The technico-sexual division of labour was reinforced by the practice of part-timing – which Home-Tec call 'key-timing'. Key-time involves investing labour in particularly busy times of the week, especially Saturdays. Key-time jobs, while they do attract some young men, were clearly designed for women of a certain age. Because of the close link between motivated selling and high turnover, a *need* to earn was the most important criterion in selecting from applicants. A senior manager's description of the ideal key-time candidate for Bunnett's stores was a divorced woman with school age children and 'a bit of go in her', whose responsibilities would limit her working hours while at the same time intensifying her need for money. Both the company and the key-timers had a problem here, however, since women with pressing domestic commitments had little chance to follow the training schedules and improve their selling skills.

The experience of one young woman, Tracy, is symptomatic of sexual divisions in the selling of microwaves and other white and brown goods. Tracy trained as a nursery nurse and worked as a nanny, then as a cashier in Tesco's. When we interviewed her she had been working as a sales assistant at Home-Tec for two years. She hadn't wanted ordinary shop work, the technical nature of the products here seemed more interesting. 'I've learned a lot and I enjoy it.' When Tracy first started she had aspirations to get into management. What changed her mind? Meeting a young man, Andy, and marrying him. Now she felt, 'to be quite honest, I'm not interested in going any higher. I'm happy where I am.' Anyway, she could see no women higher than herself, and 'it's all bloke-oriented'. If she allowed herself to dream at all, she liked to imagine herself a car-driving sales rep for some other firm. But she realized that 'they always expect you to be a lot older and, to be quite honest, I think I'll probably have kids by the time I'm that age'. Besides Andy 'wouldn't let me . . . not unless we were scraping the barrel'. Tracy was scaling down her aspirations because she had had two miscarriages while working at Home-Tec and 'they reckon it's to do with the stress, strain, and all that, you know'.

A combination of the physical realities of being a woman and the social pressures of being a wife, in the context of the oppressive gender relations of work and of home, seemed now to conspire to keep Tracy fixed where she was – and on the several occasions we met she seemed angry, if still joky. One might contrast her with one of the male sales staff, Bruce, and see the pair, as we saw Carol and Don in Electro, as symptomatic of the gender relations of their workplace. Although on the same grade as Tracy and other women he was perceived by both management and colleagues to have more authority. He was given more responsibilities, and the staff spontaneously referred to him as 'floor manager', although this was not an official grade or status. This was connected to training he had received at London head office. Bruce was clearly marked for management. On three occasions while we were speaking to the women, male sales staff intervened sometimes irritated, sometimes joking, sceptical that they could have anything of significance to tell us.

None of these gendered phenomena was an expression of formal Home-Tec policy. Unlike Electro UK, the company had an Equal Opportunities policy set out in the staff handbook and given a mention in the annual report. This pledged not to discriminate against anyone applying for a job or in their employment 'for reasons of sex, marital status, creed, colour, race, nationality, ethnic or national origin, religious belief, political opinion or disablement'. There were further statements supporting the career development of women and ethnic minorities as well as recognizing the value of older people who keep their skills updated. Additionally there was a policy on sexual harassment: 'no employee should be subject to unsolicited and unwelcome overtures or conduct, either verbal or physical'. There was a handful of women higher in the organization than any had yet climbed in Electro UK. Yet, when we asked a group of saleswomen what they felt about the company as an employer of women it drew a sardonic laugh. It was clear that a combination of differences and inequalities between the sexes outside Home-Tec and in the culture of the firm itself was producing a slippage between policy and practice.

The result was a topography similar to that of Electro, modified only by Electro's proximity to the manufacturing 'coalface', Home-Tec's situation in the commercial world, interfacing with private life. Men in Home-Tec again peopled the upper reaches of the landscape, women were populous below and thinned out progressively with the altitude. Again too there were horizontal divisions, most noticeable down on the levels, with women clustered in some kinds of job (clerical, part-time selling of white goods), men in others (warehousing, computers, brown goods sales). One fault line was again technological knowledge and know-how. Of course, selling leisure and domestic equipment did not require an engineer's knowledge. 'People expect us to be knowledgeable but they don't expect us to be experts. We employ engineers and people like that.' On the other hand familiarity and confidence in sourcing, demonstrating and explaining those artifacts considered most 'technical' was what characterized men's roles in Home-Tec.

In this area of the microwave-world, Home-Tec is linked to other actors where likewise men are the ones with technological knowledge and jobs. In the subsidiary company, Tec-Care plc, in which microwaves and other goods are tested on Home-Tec's behalf for quality and safety, the

management was male; and of the 15 staff in its test laboratory nine were male engineers, six female clerical staff. The same rationale applied as at Home-Tec. The manager in charge of the test lab said:

> we test everything from a hair-drier through to a big washing machine . . . you need quite a big background for that type of thing . . . Most with that background is people like myself out of industry, or engineers from other test laboratories . . . There's not that many women in engineering, fullstop, and to get one with the specializations I would require is quite unlikely.

After-Care, the company contracted to carry out after-sales service of microwaves and other white goods for Home-Tec, also had a male management and team of (almost all) male service engineers. The exception was a woman engineer inadvertently acquired when taking over another firm on whose staff she already was. Though it might have been expected that women engineers would be particularly welcome to women clients in place of male callers in the home, a different rationale was expressed to justify the male team. One had to consider

> the physical side of the job. There is the thing of wandering around on your own and going into houses on your own . . . especially in the winter when it's dark at four o'clock and you're going up these country lanes. It isn't an ideal situation for ladies.

Curiously, though, any sexuality-related problem in conducting home visits was represented not as a danger for women engineers but for men: 'a man may feel threatened that he could be maligned by a woman he encounters on his own'.

The second major fault line in Home-Tec occupations (again similar to that found in Electro) was an orientation to family, home, cooking and other kinds of domestic work, to people and to feelings. The reasoning was clear in people's representations of women. If masculine gender identity was constituted as being 'good at' technological things, feminine identity was found in being 'good for' tasks ranging from personnel management and training through to the sympathetic and understanding approach to sales. A divergence between an active, effective masculinity and a maintaining, sustaining and relational femininity is beginning to be clear.

Different Locations, Unequal Outcomes

The gender topography observable inside Electro UK and Home-Tec is partly (though not only) related to the predisposing patterns of the sex/gender system of the wider world, particularly in domestic and private life. Cecilia, a successful woman manager in Home-Tec, described very lucidly how, while there was no overt discrimination against women getting on in, say, marketing and buying, 'you have to be ruthless', 'you have to play a political game' in a way that many women dislike. She found that women applying for jobs at Home-Tec had much better qualifications than men: 'the men just don't match'. Women were keen and determined to succeed. Nonetheless they did not get through to the upper ranks in anything like the proportions of men. One reason, at her level as at the bottom of the tree, was the double shift. Cecilia pointed out that there were very few people high up who did not work excessively hard and for whom indeed work did not seem to be their whole world. She said, 'It seems to me if you can only progress by working 12 hours a day, six days a week, then sod it . . . I have other priorities in life.'

Down in the lower reaches of the Home-Tec landscape, the irrefutable demands of private and domestic life also shaped the attitude and achievements of Tracy, who said:

> I don't think myself there's many women who want to do that kind of day . . . A woman doesn't always stay single. A man, even though they are living with someone, going out with someone, married to someone, they can always still be [*pause*] more *singular* than a woman, can't they. Whereas us, well we've gotta [*points to her appearance*] be perfect, you've gotta be doing things. [*Question*: You've got other responsibilities?] Yeah, yeah. Like at half past five at night I feel all I want to do is go home and fall down and have a sleep. But I haven't . . . I've to go and sort things out, I've gotta cook dinner, clean up, and then *he* goes off to work, you know. That's my situation, and a lot are the same.

At this point of course the women and men who are the workers in the microwave actor-world coincide with that other highly relevant population at its periphery: the customers who buy microwaves, take them home and become their users. Many of the employees of Electro, Home-Tec and other organizations in which we interviewed were microwave owners and talked to us about that other, domestic relationship they themselves had to the oven. Often the women working at the point of interaction with the client, as sales assistants or customer relations managers for instance, said they thought of the customer as 'someone like me'. We also carried out participant observation of microwave purchase in the shops and interviewed in 20 households. We found, as might be expected, that the gendered relations of domestic life that shaped workers' relations to work (like that of Cecilia and Tracy) also shaped up a sex difference in the practices of buying and use of the microwave.

Because of the private nature of the home and the transitory nature of the moment of purchase it is not easy to quantify the sex differences. What happens and what is said to happen, topography and rationale, become blurred. We found, for instance, that when women and men were both involved in buying a microwave oven (by no means always the case) women were more likely to be the ones to express interest in and concern about the cooking benefits of the different models, men with their technological features and price. As users we found that men more

frequently used the oven in the simple mode of 'pie warmer'. If anyone did any serious cooking in a microwave it was more often a woman. We found that, in households involving a woman and man living as a couple, even where men used the oven to heat food, women were more likely to take responsibility for its provision and preparation. The interactive, systemic, nature of the gendered relations of public and private life, work and home, are clearly seen here, since the higher earning power of men, their longer paid work hours and their work-generated knowledge of and confidence with technology all have a bearing on how women and men relate to the microwave oven in domestic life. We shall see much more of these interactions in Chapter 5.

Here, to conclude our mapping of the gendered topography of the microwave-world we can now ask: what practical effects does it have for its women and men as distinct groups? We can distinguish a making of *difference* (we return to this in Chapter 6) and, through the difference, of *inequality*.

Perhaps the most obvious inequality has to do with *money*. Take Electro, where an operative (usually a woman) earned £6000 per annum and a technician, foreperson or junior engineer (rarely a woman) could earn from £8000 to £10,500 a year. Remembering women's absence from management, compare again the pittance of those operators with the annual pay of Electro's senior British management (£25–30,000) and the

Japanese UK Managing Director (£105,000). At Home-Tec an assistant manager's pay was between two and four times that of a junior sales assistant or clerical worker (characteristically female), that of a full manager from three to seven times as much. In After-Care a clerk (female), an engineer (male) and a manager (male) were paid in a ratio of 7:11:14. Women's relative absence from technical and management posts is costing them dearly.

The presence of women as the majority of part-time workers in Home-Tec is a reminder that domestic relations and work relations interact to produce women's low earnings. It is domestic commitments in the main that cause women to choose part-time work, and part-time work produces little reward. Conversely, of course, men's higher pay makes them, in the family context, relatively better off and gives them more control of money. This in turn enhances men's say in the financial decisions concerning major purchases, including purchases of domestic equipment such as a microwave oven.

A second inequality-generating difference is the nature and extent of *transferable knowledge and skills* women and men stand to gain through work. In Electro, all newcomers received a three-day induction training, including operatives, who were enabled gradually to increase the range of tools on which they had competence. True technical training for City & Guilds qualifications however, involving time off and payment of college fees by the firm, required managerial approval and depended on the training budget. Ten per cent of the workforce were currently sponsored on such external courses. Ninety-five per cent of the people sponsored were men.

In Home-Tec all sales staff were required to pass through the retail skills training process, and a genuine intention existed to ensure that women and men alike gained both brown and white goods product knowledge. Part-timers, however, in which category most women are found and the great majority of whom are women, simply did not have the necessary hours at work to enable them to complete the course. Although some were known to take the training videos and books home for free-time study, most of those missing out on training were women.

Women are coming into contact with the technological *artifact* that is the focus of this actor-world, and with the technological *processes* involved in its production, from the assembly line to the computer systems that today control everything from parts supply to sales. Yet the prevailing gender relations preclude them, in most cases, from obtaining technological *knowledge*. They were missing out on the training for technician and engineering qualifications that could carry them up and out of their non-technical jobs, to the rewards of progressive technological careers in these and other organizations. As a result besides (a point we shall return to) women wield little influence over the design and development of new technologies. Nor, despite their continuing role as managers of households, are women obtaining the experience and qualifications that are recognized as those fitting people for senior management in employment.

What other sex differences can we detect? Women and men incur different *pains and injuries* from their work. Men are more liable to strain from lifting, for instance, women to strain from repetitive tasks. Whether suffering in different ways can be classed as an inequality is questionable. What certainly does lead to a disadvantage on the part of women is the observable difference in *hours of labour and leisure*. We can assume (as we will have occasion to discuss further in Chapter 5) that women continue to do considerably more unpaid domestic labour and caring work than men. When women and men work similar paid hours therefore the overall burden of work to women is greater, and men have more leisure time to dispose of as they please. Women doing paid work part-time is no indicator of enhanced free time: it is often chosen to accommodate particularly heavy domestic commitments, such as the care of young children or the elderly, or even to fit in with a second part-time job.

A third observable pattern is differential *mobility*. We saw how, in the Electro factory, women often sat in one place, men stood or moved freely around. This reflects, in microcosm, the greater mobility of men generally. More men than women in Electro UK caught the intercontinental jet to visit head office in Japan. More men than women in Home-Tec were in jobs, such as those of branch managers, requiring a move of domicile. An interesting exception is the job of microwave 'consultant' done by women, part-time, in Electro UK. These women must travel from shop to shop, carrying out demonstrations and running courses. In case one should think this placed them on a level of equality with salesmen however, it was pointed out to us by one woman that salesmen were paid 50 per cent more than the consultants and had more powerful company cars. 'What really makes you laugh is that the males get big Cavaliers, you know, and the women get little Astras. And that,' she added, 'has got to be a male and female phenomenon, really it has to be.'

Perhaps the most important practical difference between women and men in this actor-world is simply the degree of *authority* they wield. Technological know-how gives authority over materials, artifacts and work processes. Managerial position gives authority over people and organizations. If one could simply add up the number of people in the microwave-world accountable to a woman, and the number accountable to a man, it would exemplify a great gulf between the relative sway of women and men. Men are almost always those that have the authority to allocate another person to her or his position in the topography: to make an appointment or a promotion to this job or that. Men are, not always but in the great majority of cases, the experts, the ones who know, those who are responsible, the bosses. Where women have an expertise and a responsibility it is often in a sphere that is itself given less

value than the sphere a man controls: personnel rather than line management, training rather than buying, cooking rather than engineering, white goods rather than brown. The white goods buyer in Wonderworld showed how almost literally men are associated with greater worth than women:

> For [the buyers of] the lower value items the pay tends to be lower and it's a training ground. But very few women come through the system and that's because they go off and do whatever they do . . . It tends to be the men who move through, going from the smaller items, the lower value, to the higher items, higher value.

Here we are approaching the next step in this analysis. Having mapped this aspect of micro-structure, the gender pattern in the practical location of the sexes in relation to the material phenomena of jobs, rewards and skills in the microwave world, we can move on to analyse in greater detail the way people *represent* the masculine and feminine. For the change in the gender order that can be worked by individual women and men escaping from their gendered locations is limited by the continual reproduction of the masculine and feminine as symbolic spheres, unequally valued. The next three chapters show something of the way technology is a factor in shaping this hierarchical relation.

3

The Engineer and the Home Economist

The aim of this research was to better understand the processes in which gender relations shape technology and technology relations shape gender. Mapping the gender pattern of location – the sexual division of labour and activity, together with the division of associated assets, such as leisure and money – was a first step in seeing the gender relations of technology. Numbers, however, of women and men in this role and that, take us only so far. The gendering of technology is also, and arguably more, a matter of symbolic, representational practices.

Although it is important to be clear, as we were in the preceding chapter, that there are very material differences and inequalities involved in women's and men's different situation in relation to technology, for most purposes it is impossible to separate the material from the meaning people make of it. If ten out of ten engineers in a given enterprise are men, that is a material fact with a material bearing on both sexes. At the same time, to anyone who sees or lives this situation it *tells* something, indeed it speaks volumes. It says: engineers are men, are meant to be men. Conversely, the meanings people make and the things they say actively shape material practices. When a manager says 'women aren't suited to be technicians', women who hear him may well be deterred from seeking training for technician jobs and the practical result is that technical work remains the work of men. We should therefore make no very sharp distinction between the material and the meaning people make of it.

This interaction of the material and the symbolic is the reason why we cannot say simply that having a woman join the male engineers would necessarily change the design of technology. Securing that only nine out of ten as compared with ten out of ten of the design engineers are male will not necessarily introduce a 10 per cent feminine influence, observable in the design of the artifact. Yet the more women engineers, the more likely it is that the technological outcome will be different. It is not the case that there is no connection at all between the gendered topography of jobs and the gender of resulting technological artifacts and processes. (And certainly the fact of women doing engineering tends to modify both individual and group gender identities.) What is of greater influence is the existence of the masculine and feminine as symbolic spheres, into which people and activities are located in thought and speech. A woman becoming an engineer steps into a masculine symbolic sphere and also, more practically, into a masculine professional and organizational culture. She is not a free feminine, let alone feminist, agent, even if she sees herself as such. And by no means all women would see themselves as either.

In the next three chapters we explore three moments in the circuit of production of the microwave cooking process. The first instance is drawn from manufacture, the second from distribution and the third from use in the household. We focus on the social relations of these three moments to demonstrate how meanings are spun around people's experiences and practices, and how the material and the representational interrelate. Stereotypes are continually belied by events, yet they reflect, while also constituting, a material reality. A stereotype is not 'true' but it is 'real'. We will show how the meanings people make and deploy involve the creation, challenging and reproduction of two complementary symbolic spheres, the masculine and the feminine. And we will see how the feminine is, in the process of adaptive renewal, constituted as of lower value.

The present chapter is based on our experience of Electro UK and other actor organizations close to it in the microwave-world. The theme is the role and status of home economics in microwave manufacture in the context of the consumer electronics industry. In the main it draws on interviews with five home economists working in 'test kitchens' – four in Electro and one in a competitor electronics giant with a very similar structure. We also draw on interviews with, among others, Electro's engineers, its sales manager, the product manager for microwaves and its advertising agent.

Despite their virtual absence from engineering in Electro, women, as we have seen, were not without influence on the microwave project. They were playing a part in many different positions in the microwave actor-world, but most significantly in their position as home economists employed by the company in a small unit of their own: the Test Kitchen. In the Test Kitchen the gender stereotype matched reality one to one: the home economists were women. Their manager was a man. The engineers they dealt with were men.

We saw in Chapter 1 what these home economists were employed to do. First, they were required to use their cooking knowledge to assist the sales pitch. Secondly, they were to contribute to the design of oven controls and programs. Thirdly, they were to test the performance of new ovens by actually cooking food in them. Finally, their role was to inform and enrol the potential user. Essentially, then, these women were employed to serve as an *interface* between the masculine company and the end users of microwave ovens, conceived by the company as women. The Test Kitchen's interfacing role was expressed to some extent structurally. It was uneasily located between product planning and design engineering: their manager said of them, 'They are under me in product planning, but they should really be grouped with the engineers.' The senior home economist, Dorothy, had been given responsibility for the team of women sales consultants, despite the fact that she was not herself in sales. And Kelly, the home economist placed out in the factory, said, 'They haven't really decided on who is my boss – I'm just sort of a one-off department on my own.'

Interfacing involves not only making connections, and controlling them, but also holding apart. And Dorothy, Kelly and other home economists were expected, on the one hand to reach out to women, understand them, influence them and advocate for them within the company, and on the other hand to enable a clean distinction to be retained between the masculine world of engineering and the odorous feminine world of the kitchen. Dorothy, musing on her experience in two microwave manufacturing firms, said simply, 'I think I was taken on, I've often thought about this, *to keep the women out of the men's hair*'.

The relation of the company to cooking is thus curiously ambivalent. We concluded that the ambivalence was best explainable through gender and the unequal value ascribed to masculine and feminine in a much wider sphere than the company alone. In the masculine culture of Electro, and it seems of other similar companies, there are mixed feelings about the feminine sphere of women, domesticity and cookery. These feminine things have undeniable importance for business purposes, but in the organizational culture they are somehow considered 'other', as 'not engineering' and as messy, out of place.

A Problem: Understanding Women's Needs

Let's first look at the real need the company experiences to develop the knowledge of women and cookery that alone can enable the design of a saleable cooking technology. Most of the managers and engineers we spoke with in Electro visualized their customer, the microwave user, as a woman. True, at times, when seeking to emphasize what a very large market Electro had for its microwaves they might say *'anyone*, man, woman or child' could be the potential user. But when speaking unprompted they invariably spoke of the microwave cook as 'her', not 'him'. The sales manager, Jimmy, was clear about this. He reviewed the company's other products.

> Well, if it was typewriters you'd think of a woman, wouldn't you? If it's a photocopier it could be anybody . . . If it's a compact disc player, male; if it's a video recorder, male; and microwave oven – female.

The company made it clear that it saw men as casual and undiscriminating microwave users. Thus, 'A woman might buy one for her son when he first sets up house, but he's more likely to be buying a Marks and Spencer's lasagne and reheating it,' said Dorothy. The technological skill and imagination of Electro's engineers and product planners was not primarily directed towards the ultimate, and ultimately cheap, 'pie warmer' into which a fellow (in Jimmy's words) can 'you know, speed, instant, convenient, bang it in, it's done'. That uninteresting project was the game of the retailers with their shoddy 'own brand' bottom-of-the-range models. Electro and other multinational firms selling under their own name were competing with each other to produce 'the ultimate cooking machine' using a battery of technological aids, separately or in combination. For complex and clever microwaves to be saleable, the manufacturers needed to identify and embody the knowledge of the informed and committed domestic cook and then enlist the enthusiasm of this cook in the microwave project.

Their conception of this user, the true microwave cook, was a woman. That is why, Jimmy said, 'we've always had a microwave kitchen operated by ladies, housewives themselves'. That is why, too, the team of part-time consultants working on behalf of the manufacturer in the retail shops is also female and comprises part-timers, supposedly *domesticated* women. In the Electricity Board shops, where the consultants often work alongside the sales team one or two days a week, housewife is set to ensnare housewife. 'They are the prime targets, the housewives,' explained the sales manager. 'They go and pay the bills, or they go and buy an electric kettle or toaster or whatever, and they see us prominently there.'

Given the low opinion the manufacturers have of retailers' sales personnel, particularly those of the multiples such as Home-Tec, it is understandable that they are concerned to inject womanly cooking skills into the selling process. A home economist explained,

> women want to hear from someone they know can also cook. They don't want a spotty-faced eighteen-year-old [lad]. . . There's no good having someone standing in front of the microwave oven, even if they know in some detail how that microwave works, if when a potential buyer comes to them and says, 'If I have a dinner party for eight, will this microwave do roast potatoes, and then my baked soufflé?' If that person's never understood what it's like to prepare six different dishes at one time they won't be able to pass on the benefits of that machine.

The manufacturer must reach out to women and sell to them using someone who will be recognized as 'like' them, someone who speaks their own language. It is thus disingenuous when managers say, as did the sales manager at Electro, that they just employ whoever applies, and that happens to be females. The fact that consultants and home economists are, materially, women, is not incidental.

Long before the range reaches the shops, however, during the cyclical processes of product planning and design, the firm must be in tune with what women want. To achieve this they need the consultants and home economists to be 'representatives' of womanhood within the firm. They need detailed cultural knowledge of the reality of women's lives. That knowledge is frail, however, in comparison with the weight of engineering knowledge in the firm, which is deployed largely in ignorance of domestic realities. We found that male design engineers, whether British or Japanese, had no very clear idea of the gender of the end user. A British engineer said he thought neither of a woman nor a man, just of 'a customer'. The reality of the customer was more a marketing concern, he felt. A Japanese engineer thought for a moment and came up helpfully with the observation that 'there are many operating buttons, the controls. And women have long nails so we have to consider that fact also. If too strong and big then maybe the nails break.' He also noted that the extractor fan should not be designed at a height to disturb women's hair styles.

Clearly, if Electro's knowledge of gendered lives and domestic needs were limited to this simplistic folklore the company's ovens would scarcely satisfy the discerning consumer. Of course, some compensatory effort could always be made later in selling and marketing to overcome the deficiencies of the artifact. Yet it seemed that little real thought about the domestic cook was invested in either advertising or selling microwave ovens. Electro had in recent years spent little on advertising its ovens. In the early growth period its adverts had, however, not singled out or acknowledged a user or use, but had adopted a strategy of invoking rather generalized British traditional values. A home economist said she felt the company's almost-all-male sales team could sell many more microwave ovens than they did. They made little effort over microwaves, they did not know how to operate them. 'You can talk to them about camcorders, not microwaves. When microwave ovens sell, they sell easily. It's not through the sales force's efforts. They don't *sell* them,' said Dorothy. Another home economist, Michelle, said of the men,

> they tend to go though for the marketing side of it, what is going to sell the oven. Whereas we seem to look at the domestic side. I know it sounds awful to say it, what woman's, you know, *domestic* – but you do, we tend, to look at the domestic side which really does in a roundabout way sell the oven more than the marketing side of it.

The market research available to the firm was not originated in-house, Electro's advertising agency told us, nor was it based on responses from consumers as to their wants and needs. Dorothy confirmed this: 'we have very little research monitoring housewives. Very little. I don't think any of the companies do.' What external information the company has is from retail, reporting what sold and what did not. It was clear that insofar as domestic cooking knowledge was being introduced into the whole microwave design process it was coming only from Electro's own women and these women were only in specific home economist or housewife-surrogate roles. They for their part felt theirs was an uphill struggle. The microwaves were, as Dorothy put it, 'designed by men, for a man, without taking into consideration the woman and her cooking'.

Technology-push, Consumer-pull

There were several factors tending to prioritize technology-as-such in Electro Corporation. First, it has been a feature of Japanese electronic developments that they have been technology-driven rather than need-led. The whole culture has tended to delight in clever new inventions and devices: the designers and their organizations in producing them, the consumer in owning them. A British advertiser working for Japanese clients put it this way:

> The Japanese [technologists] believe quite strongly, and it's understandable, that everybody will want the products that *they* want. There's this fascination with technology for its own sake. They just like to own state-of-the-art products. . . People are looking to buy a product for a year, buy the very latest, the one with the very latest gadgets and features. . . If it's something

that sounds different and appears to be different and there's a vestige of truth in it, yes, that will be marketable in Japan.

Secondly, until recently Electro UK had simply shopped for its range of products off the shelf in Japan. Electronic consumer products were rather assumed to have a universality that would make them saleable around the world. Thirdly, and more recently, there had developed a feeling in the company that in the UK at least Electro had in a way been *too* successful in selling microwaves. As market leader in the 1970s and 1980s the company had become all but synonymous with microwaves in the public mind. Of course they were satisfied with this success, it provided a sound business base, 'but it's not really the core of where they want their company to go,' explained the advertisers. The strategy they were now evolving for the UK corporation was to regain the high ground for the brand itself, so as to re-establish the company in the popular perception as a quality source of consumer electronics like its rivals. Electro after all was mainly a manufacturer on the one hand of business equipment and on the other of hi-fi and 'go-fi' (portable music systems), televisions and other leisure electronics. These two lines fitted together well. The company had considered and rejected the idea of teaming up with a white goods manufacturer. Microwave ovens, their only domestic white goods product, therefore remained isolated. The product manager for microwaves felt the future of the microwave range in Electro might depend upon the company showing interest in producing other white goods 'so that it's not just sitting on its own'.

The company's relative loss of interest in its microwaves (and it has to be remembered that it was only relative) was fostered by the safety scare described in Chapter 1, which depressed sales and which Electro felt they could do little to counteract. It has to be seen too however in terms of the unequal relationship between white and brown goods, domestic and leisure equipment. In the engineering mind, as in the popular mind both in Japan and the UK, what was exciting, interesting, forward-moving, was the latter. The microwave oven, despite being a domestic item for a largely female user, had once been, and been promotable as, an artifact at the cutting edge of technology. Now, for the moment, the camcorder was the entrancing new toy, following on the tail of Nicam videos and other such leisure 'gizmos'. There was no doubt about the gender factor in all this. The product manager, Keith, said:

I think because the industry is very much a male-dominated type industry, the female buyer's products . . . perhaps don't get the due reverence that they need, really. I think it's a very macho sort of environment that we're working in, in the electronics industry, and the high tech brown goods side of it tends to get all the glamour, as it were. . . Everyone talks about that side, and that tends to be where the leading edge always comes in.

The ambivalent relation between engineering and cooking is also expressed in a contrary trend observable in Electro. There was an uncomfortable feeling in some quarters that technology-push had gone too far. The product manager said that 'the whole consumer electronics market place has gone head over heels for technology, technology for technology's sake, rather than providing technology for the consumer's benefit'. The advertising executive, Matthew, confirmed this new perception: 'With more and more products coming in, and more and more high technology in them, people are getting increasingly confused about how to use the technology.' He was urging on the company the strategy of advertising 'only those [products] that have *benefits for people*. We would use the line that "Electro Makes Sense", on the basis that the manufacturer makes sense of technology to produce products that make sense for the consumer.'

Both men felt they were picking up on something the Japanese were themselves discovering: the value of ergonomic research, the recognition of cultural diversity and the idea that an artifact today in a competitive market should offer not only pride of possession but a materially improved lifestyle. Both also felt that the shift of emphasis from 'gizmo' to use was gendered. It meant a shift from masculine to feminine interests and they felt it to be no accident that their roles placed them closer than other men in the company to women actors, in one case to the home economists, in the other to the target user.

One way of seeing the new thinking was as a turn towards demystifying their technical artifacts, and in doing so, distinguishing groups of users, including making a sex distinction. One difference between the sexes being identified here was that men, because of some characteristic of contemporary masculinity, were more ready than women to be persuaded of the value of technology for its own sake, and more reluctant to display any ignorance about how equipment really worked. Keith thought men were hiding behind women in this sense, happy to see women expose themselves to ridicule to obtain the necessary instruction. Men too, he thought, might benefit from more user-friendly technology.

This tension between engineering and cooking, gizmo and use, was visible in the divergent directions taken by the microwave models within the total Electro microwave range. One potential direction for the oven was towards more intimate interaction between user and artifact. Against this was a thrust towards button-push, with an ever more complex response increasingly governed by the machine's inbuilt intelligence. This can be summed up as a divergence between the artifact as tool and as automaton.

From the earliest days of microwave cooking, users had complained that many foods emerged from the oven looking and tasting unappetizing. The main problem was that the food did not carbonize and 'brown' in the way we like. It was often soggy, not crisp. It was grey, not golden. Where was that irresistible aroma? Besides, because of the physics of microwave cooking, some foods (puff pastry for instance) were not microwavable at all. Yet because of its novelty the microwave found a market.

At one time selling a microwave was very, very easy [said Dorothy]. People
wanted them, they weren't interested in what they did, whether they worked,
whether they didn't work. It was a new technology. People were queueing up
and they would literally buy whatever was on the shelf. Someone would say
'Ooh! Look, I've done a jacket potato in four minutes in this wonderful
machine' . . . We are now dealing with people who I think need a little bit
more than that.

The developments in the microwave oven that occurred from around
1985 responded to the dissatisfaction many users were expressing with the
product. First a normal convection heat facility was added, but it could
not at first be used simultaneously with the magnetron. Then a grilling
element was inserted in the roof. Later, a rotisserie became available in
some models. Of course there was a cost to be paid in speed: the use of
traditional heat to add quality to the dish meant slowing down the cooking process, though for most purposes the oven was still twice as fast as
a conventional oven. At first there was also a cost in terms of size, but
subsequent developments refined the design of these 'combination ovens'
as they came to be called, and today quite trim versions are available at
reasonable prices. 'For the first time, we've got a model which offers all
of the traditional facilities with microwave in isolation or in combination,' said Keith.

Of course Electro was prompted to the development of 'combis' by the
attraction of greater value-added, stealing a march on the competition
and other such commercial considerations. For all that, Jimmy, the sales
manager was not wrong when he said, 'it's the woman who has been
pressing for further developments'. The arrival of combination models
coincided with the creation of the Test Kitchen in Electro UK and the
appointment of home economists. Looking back, their chief (Keith,
microwave product manager) said, 'It was in 1987 that things began to
change quite quickly in terms of the input that we had as a product planning department, in terms of the overall design and the bits that go with
the oven.' The Test Kitchen became, he said, 'more a fundamental part
of the process'.

It came naturally to the female home economists to think about the
artifact in terms of *food*. Dorothy said that she had been in the
microwave business many years and 'I feel sometimes we have just
galloped and galloped and galloped so far – *technically* – that we've
forgotten to ask the woman or cook what they want.' And she went on:

you have to put the pleasure back. I think this is why we have to go back . . .
You can't say there's any pleasure in producing a jacket potato with baked
beans on the top . . . We actually took away the skills and the pleasure that
go back for generations that are in a house . . . I think even though we have
modern women there's still that tradition that we still take pleasure in feeding
[she hesitated a moment in formulating a thought perhaps not sufficiently
'modern'] – yes, feeding your man!

Keith supported this return to tradition. The company had to 'consider their consumers much more than perhaps they ever did'. He said people might well value the convenience of rapid snacks and reheated leftovers but they still wanted 'proper meals' too.

> The Sunday lunch for instance is still quite an institution within the UK . . . getting together and eating as a family. So for us to actually sell more expensive models we've got to nurture the cooking aspect. Retain the convenience aspect, but nurture the sophisticated side to it as well. Hopefully being able to come up with a machine that is idiot-proof to use and it will give you great results as well. So we can give you Cordon Bleu cookery but you don't have to be a Cordon Bleu cook.

Meantime, a somewhat contrary set of developments was occurring in the design engineering field within Electro and other firms manufacturing microwave ovens. If the combination of traditional and novel forms of heating was a development that began from the *food*, there were other innovations in the wider world of electronics pushing for attention. Fuzzy logic, for instance, was very much in the air. Engineers tended to begin their thinking about the direction of microwave by wondering how these innovations might be applied.

The thrust from the engineering side was to greater and greater *automaticity*: the costlier and newer microwave ovens began to be provided with electronic substitutes for thought and attention in cooking. There were soon sensors that could measure the 'doneness' of food by detecting the steam emitted. There were other devices that could measure the weight, viscosity or colour of the food. The oven was able now to signal and switch itself off when it sensed the food to be ready. There were programs that required the user only to tell the oven the kind of the food and they would weigh it and take the oven through a series of cooking phases with different power settings and times, including 'standing' periods. These models of course have required much more refined touch-pad controls with complex liquid crystal displays.

One problem with the magnetron as a source of cooking heat is that it is impossible to alter the wattage at which it emits microwaves. Until recently, different 'power settings' on ovens could only be achieved by changing the proportion of the time during which the magnetron was active. The design engineers could offer the user power control only by programming the magnetron to alternate for differing periods between on and off. This produced unsatisfactory cooking results, since it amounted to bursts of intense radiation, capable of burning the outer surface of the food, followed by moments of inactivity. A recent development, known as 'inverter technology' goes some way to overcoming this disadvantage. Though the magnetron must still be either on or off, inverter technology greatly increases the rapidity with which it changes state, producing smooth, almost stepless increases in power with more controllable effect.

A more generally influential technological innovation in the late 1980s has been 'fuzzy logic', an aspect of artificial intelligence (AI).

Conventional computing operates by the logic of binary codes. It is limited to 'on' and 'off' conditions, black and white. Fuzzy logic is capable of dealing with shades of grey. It can use ambiguous information and work within a much greater arena of tolerance and discretion. For example, fuzzy logic has been applied to certain hand-held video cameras with the effect that the camera distinguishes between inadvertent tilting and changes of angle that are probably intentional. In other words, this is clever software that makes intelligent guesses. Applied to microwave ovens, 'fuzzy' will be able to take information from the various sensors and continually re-estimate cooking time as cooking progresses, if necessary overriding any program chosen by the user.

Representing the Consumer, Creating the Consumer

Fuzzy logic and inverter technology, together with more advanced sensors, will turn microwaves once again into technologically interesting, leading-edge products. The engineers speak enthusiastically about the coming developments. They present the marketers with a number of problems, however, and it is interesting to hear them talk through the contradictions, which are laden with gender significance.

The first, and perhaps the simplest, problem is how to recover the increased costs of production. Anticipating the arrival of the 'ultimate cooking machine' in a couple of years' time, managers explained that meanwhile the company was 'maintaining some realistic price points at the high end of the range', psychologically important in preparing the public for the more expensive model to come.

Second, and more difficult, how were the splendours of the new technology to be explained to the consumer in such a way that they would seem value for money? For one thing, they had misled the consumer into thinking s/he already *had* power control. 'I don't think any of us have actually figured out how we're going to do that,' said Keith, the product manager. Japanese managers were less concerned about this than British managers. In Japan more people – especially more men – are computer literate. There, new technology sells itself. Already, said a Japanese engineer, 'in Japan everything you buy has got "fuzzy" on it, and that is enough to make it a must'. Men just love 'the sheer, sheer magnificence of it all'. By contrast the British consumer is technically ignorant. A Consumers' Association survey found that scarcely any respondents had heard of 'fuzzy logic'. The British are believed to be resistant to such gimmicks. 'Someone who wants a sponge cake,' said Marion, a home economist, 'they don't want to buy a machine because it's got fuzzy logic, they want to buy it because . . . their sponge cake will come out the way they want it.'

Some people speak of a phenomenon they call 'technofear' in the public. Women are supposed to be particularly prone to nervousness

about new technology. The problem for the marketers is that, if they use the cleverness of the machine as a selling ploy, consumers will be at best unmoved and at worst put off. The answer might be, it was thought, to raise the issue of 'technofear' head-on in promotional material, introducing the new technology as antidote along the lines 'the more technology you apply the less you need to worry your head about it'. This at least would enable the intelligence of the machine to be made public, and so legitimize the high price. It would run the risk however that sales would be killed by technofear or techno-indifference. Besides, 'People don't want to be told they're stupid. We're going to have to find a way of actually expressing that.'

An alternative was to stress the benefit of automatism without stressing how it was achieved. That benefit would be defined as 'user-friendliness', because basically fuzzy logic in the kitchen 'enables complete novices to produce quite sophisticated end-results by pushing two pads'. The home economists however could see that this business of instant clever cooking by an imagined user who is both technologically uninformed and a *non*-Cordon Bleu cook ran counter to their best instincts, mobilized around the combination oven, to *endorse* the traditional skills of the domestic cook. In the short run, at least, the sensory facilities could not cope with combination (dual-heat source) cooking. Pure microwaving was implied. A discerning, all-round cook would thus scarcely be satisfied by it. On the other hand, its wonders would be over the head of the average pie-warmer. It sometimes seemed as if there really was no obvious consumer who could be considered the target for the highly automatic programmable microwave oven.

Selling to a domestic cook, the home economists knew, was different from selling to an amateur photographer or a music fan. Marion, Senior Home Economist in one of Electro's rival firms, put her finger on the difference between leisure electronics and white goods in the following exchange.

> *Marion:* In cooking, you've got a person doing what they do to it *before* it goes into the oven. All the oven does is do something to the thing once it's gone in. With a camcorder, if the child refuses to smile, it's not the camera operator's fault. You've got to take into account what she or he did to the food before it went into the oven.
>
> *Interviewer:* It's work?
>
> *Marion:* Yes, it's a labour process and pre-existing skills that are being used in conjunction with the oven, which is not the case with a camcorder or a hi-fi set.
>
> *Interviewer:* Is it also that it's something done for other people?
>
> *Marion:* Yes. The music is not done for somebody else, it's for the person who switches it on.

It scarcely mattered, she felt, if the purchaser of a camcorder or hi-fi system, more likely to be male than female, were attracted by the hard sell of technological wizardry while actually failing to understand half the functions he was buying.

But somebody buying a cooker, at the end of the day, they want to get a meal on the table. If that meal is inedible it's a reflection on them. They will try and blame the oven, but the husband and the family who don't like it will blame *them*. So as part of their confidence, that oven must *work* for them. So it's a very different thing. Anyone can stick a record or a CD on the player, and if they don't use all the features it's capable of no-one's going to know. The sound's still coming out. But food, if it's inedible. . .

The skills and labour of the traditional cook, they seemed to be saying, called for a response from oven manufacturers that would produce a tool to assist and enhance their work, not an automaton to substitute for the cook – which would in any case be bound to produce inferior results. They suspected that the new developments were yet more evidence of the engineers' technological entrancement. 'We make the technological features, and then we make the housewife desire them,' said Dorothy.

This makes the work of the home economists highly contradictory. As interface between producer and consumer (and as indeed personifying these two actors since home economists are not only employees but also 'housewives'), they are pulled in two directions. As 'good cooks' they tend to be sceptical about simple microwaving. Yet they must promote microwave-only models and indeed spend much of their time formulating self-cook programs for them. They wish to defend and amplify women's traditional cooking skills, and yet are obliged to render them in some sense redundant. They are employed to represent and 'stand for' the domestic cook, yet must defend the manufacturer against criticisms of microwave ovens that arise from misuse.

One aspect of the job involves looking inward, feeding traditional cooking knowledge into the artifact in the form of control panels and cooking programs. One young home economist at Electro, Helen, expressed the frustration involved. Very often the programs for which she prepared the data, based on oven tests and given constants and formulas, failed to produce properly cooked food in practice. Cakes for instance sometimes surprised her by over-cooking.

> There are so many differences in the way somebody makes a cake, and in the ingredients . . . When it says auto-cook, this is what all my testing is about, it should be able to get it spot on, no excuses. [But] there are always going to be eggs that are different sizes to the ones I've used, or they might've put in the wrong ingredients or the wrong quantity. That's always going to crop up. . . [Besides] everybody has their own likes and dislikes, and what's over-done to me would be just satisfactorily cooked to somebody else.

The problem, of course, is that cooking involves a great deal of tacit knowledge. A practised cook, let's say a housewife of many years' standing, uses not just a printed recipe but the memory of many past successes and failures. To tell when the dish is ready, she uses not only a clock, but all her senses. She listens for the bubbling, sizzling or spitting. She uses her nose to tell if burning is imminent. She judges the colour by eye, and the viscosity by touch. She uses her knowledge of food products to predict their behaviour in cooking – the difference, for instance, between strong flour and refined flour, between mild and matured cheese. She uses nutritional knowledge: she knows if a recipe contains an unhealthy amount of fat or salt, and adapts the recipe (and cooking time) accordingly. She also uses social knowledge: who in her household likes food well done, who prefers it under-done; who likes it sweet, who likes it tart. Most engineers do not possess this kind of knowledge and, though they may suspect it exists, they seldom use it as the starting point for design, which begins instead with recent advances in technology.

This gulf between the technology of engineering and the knowledge, which should also rightly be called a technology, of cooking, gives its urgency to the home economists' second role. If microwave ovens as artifacts do not immediately fit domestic cooking practices, if their use is not immediately transparent to the purchaser, some medium for *educating* the cook is called for. Essentially it is the task of the home economists to help bring into being a new kind of domestic cook interested in microwaving and appropriately educated, a purpose-designed customer adequate to the range of ovens being marketed. The means available to them are scarcely sufficient: the customer help-line, the instruction manuals and cookbooks she writes for inclusion with the microwaves.

The problem of communication between manufacturer and user of microwaves arises in particularly acute form when health and hazard scares arise. On the one hand, if users were to follow the manufacturer's instructions to the letter most accidents would never occur. If ovens are kept clean there is less danger of fire. If cooking times and guidance on cooking container materials are observed the likelihood of explosions or arc-ing is minimized. If the injunction to allow for 'standing time' is followed food is more likely to be safely heated throughout. Despite their loyalty to the domestic cooks, the home economists often expressed exasperation that they 'don't read the instructions', 'don't refer to the booklet'.

The frustration of the home economists arises in part from a contradiction: they both need the microwave user to observe instructions *and* need her to use traditional cooking common sense. The best and safest microwave cooking results not from slavish adherence to rules but from intelligent interpretation. Nancy, a home economics teacher in a technical college, balancing the share of blame in the listeria poisoning cases said:

> I think it's this kind of obeying instructions because it's a different technique. The senses aren't involved in the same way with microwave cooking as they are in normal cooking. . . A minute is a long time in microwave cooking. In a microwave the type of timing that you're doing has to be more precise. It's crucial and critical in a way that it's not with conventional cooking.

Time and again you hear microwave home economists insisting in this way: 'this is a *different* kind of cooking, you must follow our instructions'. Yet they can also be heard arguing the contrary. Nancy for instance also said:

> it seems to me people are just leaving their common sense at the door. They're not applying normal common-sense principles. If they're prepared to eat a piece of chicken that's partly raw, well, would they be prepared to do that using a conventional oven?

The home economists' work on auto-programs aims precisely for mindless cooking. Yet, Helen too complained that 'the consumer seems to think they can just press a button and it can be cooked when it comes out'. She went on:

> they seem to have this theory whereby the microwave oven does it all for them. Using a conventional oven you are constantly opening the oven to see whether it's cooked. Whereas with the microwave it seems to be you shut the door, press a button, and don't open it or look and it until the time's up and the little beep comes up.

Likewise Marion said too many consumers were resistant to opening the door and 'poking around with a knife in the way you would in a saucepan'.

Another aspect of the contradiction is expressed as a generation gap. The microwave manufacturers' home economists should perhaps logically welcome, as the natural clients for microwaving, a new young generation of domestic cooks with no preconceptions about cooking. After all, they are themselves changing the practice and language of cooking. A traditional cook knew what was meant by 'roast' and 'bake'. Both were done in an oven, but the first term applied to meat, the second to cakes and bread. By contrast, in a microwave program the terms relate to control settings, 'roast' may be using medium microwave power combined with 250-degree Celsius convection, while 'bake' is 200 degrees with medium-low microwave power. Yet the home economists tend rather to deplore the dying out of traditional skills. Nancy said:

> a lot of the things about food, you don't even know you're learning it as you take it in, do you. You just see what your mother does and copy it. . . So many kinds of old-fashioned knowledge – younger people in their twenties now – there's a big gap. They don't have that knowledge. It's still needed, but it's not there.

The home economists' interface role is thus a frustrating one. When the press exposes microwave hazards the home economists, on the defensive, scarcely know whether to blame the consumer watchdogs, the media, an ineducable public or themselves for their failure to communicate proper microwave cooking practice.

Relative Value: Engineering and Cooking

Ambivalence – in the sense of ambiguous value – runs right through the
experience of these home economists in the engineering field. On the one
hand they are undoubtedly accorded a certain value, on the other they are
certainly undervalued.

As the only people with cooking knowledge in the enterprise they do have
a degree of autonomy. Though they have bosses, the bosses know little of
their specialism. Their very location in the white goods area enhances their
ability to influence events, because Japan has been readier to devolve
design to its British engineers in the case of white goods than of brown.
Marion, Senior Home Economist in one of Electro's rival firms, said:

> the marketing department that cover the brown goods, they're all male except
> for a few women who are in administrative roles, and they probably do
> believe, yes, that they're of some bigger importance to [the company]. But if
> you look at our department, we actually have more of a share in the design
> of the product. Brown goods technology comes direct from Japan, and all that
> department do is to provide advertising and sales materials. White goods,
> which includes personal hygiene, vacuum cleaners, bread-makers and
> microwave ovens, we're bigger, we have far more women.

She went on to explain that nobody in the firm could substitute her
knowledge.

> I'm not saying I haven't had to fight to get a high profile. [But] because . . .
> we have a point of view we can substantiate, we've been able to keep that
> profile up. My English marketing manager knows that if he goes to look at
> some new products in Japan, he would not be able to evaluate them without
> my knowledge. He won't know what a new heating system is going to do to
> a cup-cake. We have *information power*.

The home economists at Electro UK confirmed this. Michelle gave an
example. The engineers had planned to make Yorkshire pudding the
subject of a program on a certain oven, and had to change their tack
when she informed them that this dish could not be cooked by
microwave. 'Do they listen to you?' we asked Michelle. 'They have to.'
Dorothy turned to her. 'Japan listen to you. . . You have the final say,
don't you?' 'Well, yeah,' Michelle affirmed cautiously, 'you do actually
have the final say.' Beyond the walls of the Product Planning department however it was suggested that any difference home economists were
permitted to make to microwave oven design was cosmetic. Matthew, an
executive in the company's advertising agency, said, 'I'm not sure that
a woman has ever been in with the engineers as part of that process. I
just sense that the whole process doesn't happen in Electro.' He believed
the home economists' role to be confined to product demonstrations,
retail support. It was as though he described an interfacing mechanism
limited to a one-way effect: easing the movement of the artifact out to
its public, unable to feed back effectively information that would truly
shape the artifact.

The outward function, however, was unquestioned. These home economists, internal to the manufacturing firms, at the heart of the microwave actor-world, have their own network. 'We discuss the role we play. We discuss the problems we have, how we'll solve these problems. There are very few secrets.' They are also part of the bigger community of professional home economists, some of whom, as publicists, teachers and writers, as well as employees of related sectors such as the food and container industries, have been the single most important group of actors in enrolling the cooking-and-eating public in the microwave project. In doing so, they have, it will be apparent, translated the project according to their own interests – emphasizing food quality, de-emphasizing engineering tricks.

Despite their undeniable importance to the manufacturers, the home economists are relatively undervalued. As we saw in Chapter 2, they are, considering the responsibility they have, under-paid. Kelly said, 'I don't think they appreciate the skills I have, sometimes. I think a lot of people think "she's just sitting in a kitchen cooking". I don't think they realize the importance of some of the work I have to do. . . Even on the pay scales. I'm not paid on the same level as the engineers, even though I have a degree the same as them.' She was in fact paid around one-third less.

It was not only a question of pay. The way their work was situated and represented showed the low value accorded it. 'If it was any other company and I was working with any other product, I would be working in "new product development" or something like that, *creative* product development,' said Helen. As it was, the home economists in Electro UK felt they were seen as working less in a laboratory than in a pantry. They had had a tussle to get the words 'test and development' put in front of the word 'kitchen' on the sign on their door.

If they were not rewarded with money, they likewise were not recognized with status. They were not given access to the external reputation they could have won had they been permitted to be the named authors of their cookery books. Nor were they accorded internal standing. Dorothy, the Senior Home Economist at Electro, was not technically graded as a manager, though she was responsible in practical terms for the running not only of the Test Kitchen but also the team of consultants. Helen could see no promotion in sight unless she transferred to the marketing department. And Marion, in a different company, was having a similar experience. She could not, she said, honestly complain about her salary. However, she went on, 'though I very much enjoy my job, I know I've been abused. Because I'm very conscientious, they've taken it for granted.' She said if her boss, the marketing director, were to drop dead tomorrow she 'wouldn't even be invited to apply for his position'. For one thing, 'as in any company, when you're in a job and you're doing it well, they're very happy to leave you there'. For another, 'to move, we would have to leave behind a large chunk of our expertise'.

Cooking expertise, unlike engineering expertise, was not considered relevant to management in an electronics firm.

The women were all aware that the reason they were under-paid and undervalued was because they were seen, as Kelly says, as *cooks*. Indeed, Helen said of Kelly, 'to a manager above, who doesn't have a direct knowledge of what she does, she's just a cook. . . which is absolutely infuriating'. And for her part Helen felt people thought of her as

> a little girl, and you're in the kitchen. . . So many people come in and say, 'Oh, d'you cook all day?' I *don't* cook all day, I wouldn't call it cooking. [Cooking] is just the function of the machine that I'm testing.

The women reasonably see themselves as doing a kind of engineering or science. Helen went on to say she would not be with the company if this were a cooking job. 'I like the *science* side of it, it's a very, very interesting side of it to me.' Dorothy was also clear this was to do with an unspoken comparison between two unequally valued occupations. The general manager, she believed, had little idea what went on in the Test Kitchen. 'It's maths, heating curves. But he thinks, "It's not high tech, is it?"' To underline the point she added:

> We have one department here that everybody is in awe of . . . that is the data processing department. Because they are working with machines and things that most of us don't know about . . . they are the whizz kids of Electro and are viewed with the greatest of respect for that.

Of course, what was happening here was not only the misconstruction of what the women do. True, they do not merely cook. More importantly though the value of cooking was being denied. Cooking itself is a technology, involves skill and should be accorded value. It was interesting to see that the one compliment we heard a home economist paid by an engineer was a back-handed one, suggesting her work was of interest precisely because it was *not* cooking. He had told a newcomer to Electro, 'Go and spend some time with Kelly next week. You'll be surprised at what she does, as well. *It's not just cooking.*'

Dorothy was also clear that the undervaluation of home economics and home economists had to do with gender. When we asked her how well she felt her skills were recognized in the company, she exclaimed, 'I wish you hadn't asked me that!' She replied thoughtfully and emphatically, 'it's [*pause*] home economists are classified as second class citizens in the company . . . very much so. They're cooks. They're *women*.' Home economists were being appointed partly because they were women (their gender was a quality the company needed in them), but were being accorded low value because they were what they were.

The gendering spilled over in interesting ways on to the microwave oven as a product within the context of Electro. We were told, by an informant who wished to be off-record, that microwave was seen as a low-status product among the sales force. Helen felt the microwave side of the business was neglected as far as promotion expenditure went.

Though she saw her cookery books as significant marketing tools, 'very very user-friendly', their production was starved of funds. Keith confirmed that domestic equipment was seen by most of his male colleagues as 'pedestrian and female'. He felt sure they treated him differently because he was involved in the microwave side. 'They would find it very difficult to raise the same levels of enthusiasm for microwaves as they do for other product groups,' he said. It related right back to the assumed female sex and domestic function of the user.

It was interesting that Marion had said, 'we're not considered like engineers . . . because we have this consumer interaction, and I still believe that as a woman we have a definite feature there!' Women were being employed to be in contact with other women, out there in the home, yet simultaneously being written down because of that association. Now, quite independently, Keith elaborated on this theme.

Keith: I think . . . business equipment falls between the two camps of the exciting, macho, brown goods world and the domestic appliance, women-oriented sector.

Interviewer: But . . . surely business appliances are quite high tech? And in some ways microwaves can be too?

Keith: Yes. I suppose it's not the actual quality, not the technology that's involved, it's the character of the product group, the persona of that particular group, and how it's perceived by the users and the purchasers of those products.

Interviewer: It's to do with use?

Keith: Yes, I'm sure. And probably situation as well, where it's used, the environment it's in.

Keith was highly self-aware, and appeared to have chosen, or at least accepted, working with a white goods product precisely because it was, in the context of the electronics industry, relatively feminine.

I've probably done it subconsciously. I like to feel that I'm champion of the ladies' cause in terms of raising the profile and the respect of the microwave department as a whole . . . for them to stop feeling that we're poor relatives to some of these other boisterous, macho groups that are about.

He connected the division of the products into gendered categories with the gendered cultures within the organization. He said he sometimes felt himself a round peg in a square hole in the world of leading-edge technology.

I know for a fact that I've always got on better with women than I have with men in terms of expression and in terms of – ladies don't *posture* as much as chaps. . . and I like their honesty and frankness. In fact, I've always worked with women, regardless of lots of jobs I've had, and I like that side of it. I'm not too good in the male arena actually. I can be one of the lads but I don't like the bluffing and counter-bluffing that goes on within the male world.

The relatively low value we see here in Electro accorded to this group of interests – women, home, food, needs – is prevalent too in the wider world. Our forays into further and higher education showed that courses have been moving away from this sphere in an attempt to increase their relative status in academic institutions and appeal to potential employers. Young people, pressured by opinion, endorse the change. Nancy, the home economics teacher, reported:

> students here hate telling people that they do home economics, because it's thought that either you do cooking or you sew, and they feel that that's a wrong perception by the general public or by their boyfriends or people on other courses about the depth of the science they do, the bio-chemistry. They just feel that it doesn't do them justice.

Consequently, home economics degree courses are taking new names: 'resource management', 'food and consumer technology', 'applied consumer science'. Another teacher explained that home economics had been about 'need and provision for food, shelter and clothing'. It had been primarily focused on the home. Graduates may well have gone to work for appliance manufacturers or the food or clothing industries, but they would have seen themselves as 'able to match need and provision, in other words to be an interface', in a positive sense, between public and private spheres. The centre of gravity had now shifted from home to enterprise, from care to technoscience, from need to consumption, from provision to profit. A women's sphere, to be taken seriously, has to divest itself of feminine associations and enter the symbolic sphere of masculinity.

Something similar appears to be happening in school education. There has long been an educational concern about the sex-segregated teaching of home economics on the one hand, woodwork and metalwork on the other. As educational philosophy shifted increasingly to coeducation, teachers asked themselves: 'Why won't boys choose the home economics option, why won't girls go into the manual crafts?' It was widely argued that content and terminology of home economics would have to change if boys were to be attracted to the option. It should emphasize, for instance, industrial design and consumer interests. The author of an article in *Home Economics* made heroic efforts to conceive of ways of adapting domestic science to, as he put it, a 'male interest input'. 'The garden city movement, including the original developments at Letchworth, can be an interesting starting point for the male students,' he imagined. Or what about 'the ergonomics of everyday life?' Or could we link the whole thing to 'organization and management' or to 'art'? (Lazell, 1981–2).

The National Curriculum introduced under the Conservative government's Education Reform Act 1988 took a more drastic step. The Department of Education's consultation document on the curriculum included technology among the 'foundation subjects' but made no mention of home economics (Department of Education and Science, 1987). Despite an outcry in the home economics profession, in the National Curriculum that ensued its themes were to be relocated mainly under 'craft, design and technology' (Department of Education and Science, 1989). Once again, a subject perceived as feminine and therefore academically inferior had been subsumed into the masculine symbolic sphere. To be modern, girls now must formulate their interests within the terms of science and technology, while boys were again let off the hook of learning about human need, techniques of care and domestic skills. Not surprisingly, when home economics and technology departments in schools have amalgamated under the National Curriculum it has been more often than not the (usually male) head of the technology department who has acquired the headship of the new department of craft, design and technology.

The superordination of engineering, the subordination of home economics, is another face of the public/private split and the denial of significance to the daily reproductive processes of the home, characteristically women's concern. The industrial world literally feeds off the private world, uses it as resource (cheap female labour) and as a market (for its consumer durables), but otherwise appears to need to keep it at a distance. This apparent fear of contamination of the masculine by the feminine has comic expression in the issue of 'cooking smells'. Marion told us of her manager's preoccupation with containing the cooking smells from her test kitchen, which tended to waft out inappropriately into the business environment of the company's smart head office block. Electro's microwave consultants, and those of other manufacturers, had more or less ceased giving cooking demonstrations in shops because, retailers had made it clear, 'the directors don't like food in the stores'. Enticing as appetizing odours are, Marion explained regretfully,

> the microwave section might be next to camcorders or television, and they don't want ordinary people walking in and smelling food cooking. They feel the shop displays are a very high-tech sort of image and it's just not appropriate to have the smell of cooking food.

4

White Goods, Brown Goods

In the previous chapter we saw that women and men as sexes are differently positioned in relation to technologies of engineering and cooking. Listening to the meanings people make of these relations told us something about technology and about gender. We could see how new technologies are shaped in, and partly by, gender relations. Technology is not innocent of gender, any more than it is innocent of class.

We could also begin to see the way technology enters into the constitution of gender difference and inequality. The notion of 'technology' is deployed in the symbolic differentiation in which the meanings of masculine and feminine are made and in the ascription of unequal importance and value to these spheres. For instance, the very selection and labelling of some phenomena rather than others as 'technology' is significant. Cooking, as much as engineering, is a technology. It involves using tools to transform matter. It is a production process. It involves special knowledge. Yet the two processes are kept conceptually quite distinct. The word 'technology' has come to mean processes men typically create, control and use, not those that women characteristically do. In this dichotomizing, engineering as a masculine technology is ascribed high value, cooking as a feminine not-technology a relatively lower value.

In this chapter we will follow the microwave oven out of the factory and into the shops, out of the hands of its originators and producers and into those of these other actors who have their own uses for it and will translate the microwave cooking project in line with their particular interests. In an electrical goods store a microwave oven becomes just one of a range of consumer durables. These are all engineered artifacts, ranging from hair-driers to satellite dishes. Yet we shall see that not all are considered equally 'technological' and those that are represented as having more 'technology' about them are considered more masculine, of more interest and of higher value.

The chapter is based mainly on the case study of the major 'multiple' we called Home-Tec, with its extensive chain of stores. This material is supported by additional interviews carried out in the offices and stores of one of Home-Tec's several competitor chains, Wonderworld. Our participant observer, Ken, also reports from his experience as a senior sales assistant in Wonderworld.

The culture of retailing is very different from that of manufacturing. There is greater fluidity – stores can open and close faster than manufacturing plants; selling techniques and advertising ploys can change more easily than production processes. The culture of customer service generates a certain emphasis on human relations, however superficial or manipulative these may be. Competent store management and sales technique involves psychology: noticing the customers as people, deducing their motives, predicting their reactions. 'Everyone who comes in that door is different from the one before them,' said a store manager. 'You can learn a tremendous amount and have a lot of fun. They'll come back and ask for you. They'll not go anywhere else.' Encouraging consumption involves creating enjoyment. 'If we can get them to come in with a smile on their face and leave with a smile on their face, albeit we've relieved them of some money in the meantime, then *we've* got a smile on our face. We're all happy.'

At one level a certain good humour spills over into the experience of employment in the retail trade. More informal interaction is allowable between shop staff, and between them and their managers in a firm like Home-Tec, than is deemed appropriate on the factory floor of a company like Electro UK. The commercial imperative of expressing sympathy with the customer to some extent favours qualities perceived as feminine. A Home-Tec area manager, John, for example, described a situation where a man had come into one of his stores to buy a camcorder in order to keep in touch with his grandchildren, who were emigrating to Australia. John said the young salesman showed little empathy with this customer about to be deprived of his family. A woman sales assistant by contrast, in particular an older woman he had in mind,

> would have been in tears and the guy would have bought thousands of quids' worth of stuff, right. . . Yeah, that's the gulf. . . If you walk in with a pram to Marion's store, she wouldn't even talk to *you*, it would be, 'Oh, what a lovely baby', you know. 'Nice pushchair.' The customer relaxes.

Yet at another level the easy-going appearance of the shop is misleading. Top management sets up competitive structures to motivate and control branch management and sales personnel. Ken described the Wonderworld form of management as being based on 'the competitive ideal'.

> Managers, and hence stores, are seen as being in competition with one another. This competition leads to bravado and rivalry. The system's reinforced by the issuing of weekly results in the form of a news sheet. This is a breakdown of the business of the previous week. The tone of the publication's 'knock-about', and yet at the same time there's a serious side. This is achieved by combining the whole of one area's results, which homogenizes the competing stores into one big happy family, which is in turn in competition with the other areas. This situation reminds me of how the lads would go over the weekend football results at school on a Monday morning.

Although the competition involves rewards, it also involves blame. Ken described monthly manager meetings as 'a chance to pillory managers who have not hit their targets'. The pressure feeds down from regional management to area management, area management to branch management, branch management to sales staff. In both companies each sales assistant has annual and monthly sales targets, and, as one young woman in Home-Tec put it, if you don't meet your targets 'you get an ass-kicking'. After leaving his temporary job in the Wonderworld store, Ken reflected

> it seems to me that it's a system which is aimed at a male competitive instinct, and the situation in my store was that the female members of staff felt, and were thought to be, in less of a challenging position. This was due to the fact that there was not one female full-timer in the store and hence none in direct competition with male members of staff.

Male Jewellery and Domestic Workhorse

A gender structure, in which 'technology' is a differentiating principle, is observable throughout Home-Tec. As mentioned in Chapter 2, the company has two distinct kinds of stores that we termed Arrow and Bunnett's, respectively, each with its own particular character. To understand the distinction Home-Tec makes between the two chains, it is necessary to perceive the major division that runs through the overall category of electrical consumer goods: the division into 'brown goods' and 'white goods'. The choice of colours for these designations only loosely reflects material fact. Brown goods are normally black, white goods are normally white or off-white but may also come in options of brown or grey. The real distinction is one of function. Brown goods are for leisure and entertainment. They include television, video recorders, music systems and cameras. White goods are for domestic work and, in the case of some small items such as hair-driers, for personal hygiene. They include washing machines and dishwashers, fridges and freezers, cookers and microwaves, vacuum cleaners, toasters and electric kettles.

Arrow stores sell only brown goods. Bunnett's stores sell both brown and white goods. Home-Tec visualizes the customer-base of their two kinds of stores quite differently, and designs its sales promotions accordingly. Sarah, a senior product manager in Home-Tec's commercial department, saw Home-Tec itself as having a highly masculine culture which was reflected in Arrow stores, which were frankly 'male stores'. A Bunnett's branch manager, Fred, said, 'They're looked upon by the powers-that-be and also by some, certain sectors of the public, as the high-tech side.' One would always expect to find the latest, state-of-the-art leisure electronics on display in an Arrow store. By contrast, Bunnett's stores are represented as 'family' stores. If the strength of Arrow lies in having the latest and the best, the strength of Bunnett's lies in having everything a family could need, 'a complete home package'. Home-Tec sees the sort of customer that typically goes into their Arrow stores as young, single, with a disposable income, and predominantly male; the typical Bunnett's customer is a couple. According to Fred:

we cater for a different type of market really. You know, the young whizz-kid type, executive sort of people, who are into computers and that, would go into [Arrow] in all probability. Whereas we get the family, you know. . . There are mums and dads shopping. . .

Fred's area manager John summed it up: 'Bunnett's is a service site. So it's like the family/home/shopping job. Whereas Arrow is high technology, the latest kick. That's the two messages.' Implicit in distinctions between Arrow and Bunnett's was a dichotomy drawn in Home-Tec between inessentials and essentials, between desires and needs, between play and work. Bunnett's was 'where you buy what is deemed to be necessary', Arrow was 'toys for the boys', 'male jewellery'.

The microwave innovation, as we established in Chapter 1, is more than the artifact alone. It also involves the carefully projected image that accompanies the product on to the market. The retailer helps shape the technology partly by designing the point-of-sale promotional material that sits with the oven in the shop and with which it is taken in, in the same glance, by the shopper. This material is consciously designed by Home-Tec with the two kinds of customer-base in mind. The company's creative services manager said that

> with [Arrow] you've got it geared up very much to the high-tech type of customer . . . they tend to be swung with gimmicks and high-tech type of gizmo-type things more than we do. I think probably the way it's looked at, it's a sweeping generalization, er, women tend to be a little bit, er, less hyped up by that type of thing.

The kind of images used in promotional material in Arrow shops would, he said, be photographs of 'someone skiing, or wind-surfing or hang-gliding'. The display cards would draw attention to anything that could reasonably be called 'new', and emphasize the 'unique selling point'. In Bunnett's by contrast imagery would be of the family, 'a bit like a holiday brochure' with 'the wife looking a tiny bit plain, maybe dowdy slightly, not too glamorous, and the two-point-five kids, you know'. They worked on the principle that 'the Bunnett's customer is deemed to be technologically dyslexic', and used 'simple, clear-to-read graphics; soft, fairly soft type of colours' on cards using icons, little pictures which sum up the features of each product in a 'very much more user-friendly way' than would be thought appropriate for Arrow customers.

In a Bunnett's store, the layout of the goods follows the brown/white principle. Ken described a Wonderworld store (similar in this respect to Bunnett's).

> There's a strict division running down the centre of the store between white and brown goods. On entering the store the customer's left in no doubt which side is the fun side and which side the boring, working appliance area.

The contrast of meaning is symbolized by the colour-coding. White seems to imply clean/simple/transparent/functional/vacuous. Brown seems to say complex/clever/obscure/challenging/contentful. Although

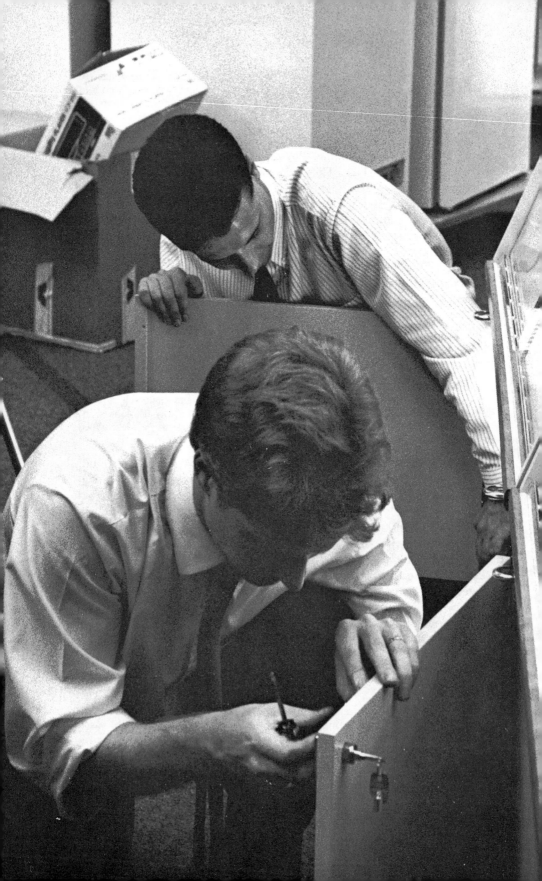

white goods are not devoid of value – indeed as consumer goods on display they are laden with potential value – they are, as we shall see, continually contrasted by the retail actors with brown goods and ascribed *relatively lower* value. It is a move similar to, indeed the same as, the dichotomizing and hierarchizing of masculine and feminine. It is customary to promote the sale of white goods through two particular benefits: cheapness and serviceability. The competitive price of white goods is frequently emphasized, and promotional placards loudly adver- tise the availability of cheap credit. The back-up service of quick response to queries, complaints and breakdowns is held to be a key attraction to the white goods shopper. Despite the fact that many domestic appliances today involve quite complex electronic controls, the complexity of domestic appliances is played down and utility is stressed. 'White goods are very much simpler, I'd think so, yes,' said Fred. 'Very much so. Take the washing machine, all our clothing has symbols on it *telling* you what to do. You don't even have to read! [*laughs*]'

A manager in the company contracted by Home-Tec to service electric consumer durables expressed the value-laden difference very clearly.

> Brown goods are looked upon as more high technology than white goods, because the white goods are workhorses, they're functional things. Brown goods, to me, perhaps you could say are more entertaining. The washing machines, the refrigerators, they've got to work for a living, they're there to do a job. . .

His technicians felt servicing brown goods in the domestic setting to be both more pleasant and more prestigious than servicing white goods.

> Brown goods is not such a dirty job. They're lighter. You're usually working in the lounge or living room and listening to the radio or television. Whereas in the kitchen, that's where the work's done. Pull it out and there's all grease down the sides, and all the rest of it. So it isn't such a nice environment to work in.

As was the case in Electro UK, the company car was both used and understood as a symbol of differential status. Brown goods engineers, this manager explained, were able to run around in cars, whereas white goods engineers normally drove a van, rationalized on the grounds that they had to carry heavy equipment. When the management tried to remove this distinction, levelling down to vans for everyone, 'it didn't go down very well' with the brown goods engineers, who felt the change 'immediately chopped them down' to size.

Further evidence of the relatively low regard for white goods was the fact that the market research effort made or commissioned by Home-Tec had all, in recent years, been directed towards understanding customer profiles and behaviour in brown goods purchase. Little seemed to be known about the white goods market. About the preferences of that consumer 'I've no idea whatsoever', admitted Cecilia, the Home-Tec business planning analyst we met in Chapter 2.

Gendering the Seller

Home-Tec management explain the sexual division of labour in the stores as being, first and foremost, a response commanded by customer expectations.

The sort of customer that typically shops for brown goods in Arrow (the Home-Tec training officer said) 'knows his stuff' and wants a salesperson with 'every little bit of technical knowledge'. On the other hand, those who shop in Bunnett's, particularly white goods shoppers, are seen as 'less demanding', more out on 'a family outing'. The sales staff and managers, as we saw in Chapter 2, tend to portray customers as associating both the technological knowledge and the authority called for in the selling of brown goods with men. Interestingly, however, there is seen to be an age factor at work here too. If 'the customer seems to expect the manager to be a male', as one store manager said, he is also expected to be of a certain age, otherwise he will not be accorded the necessary respect. Conversely, technological knowledge is associated in the customer's mind, it is said, with younger men. 'Well,' said an area manager, 'a lot of older men aren't [technologically oriented] really. If you've got to go into Arrow and ask somebody about a computer, you won't touch 'em, will you.'

The sexual division of responsibilities in the average household is seen as resulting by contrast in women having particular expertise in housework. 'I think ladies possibly can sell the white goods image better than what a man can,' said a Wonderworld manager. 'It's a ladies' domain and as such they seem to expect a lady to be able to give more information than a man can. . . A lady's more convincing.' It is a question of generating confidence and here again age and sex are articulated together. A Wonderworld branch manager, Harry, said that

> customers are happy to buy white goods from any age woman. They are reluctant to buy white goods from young men, on the basis that they think *they* know more about white goods than any young man does. A woman who specializes in white goods is seen as a specialist. On brown goods, certainly with hi-fi, they'd be more comfortable with the men.

If pandering to the customer's expectations is one reason for sex-segregation on brown/white lines, another is simply building on the sales assistant's strengths and enthusiasms. Harry said of his young men on the brown goods side,

> [they are] the type of person who comes here and has that ability and wants to do it. They learn from their peers. I can learn from *them*, often do. When a new product comes in it's like bees in a beehive. There's a buzz. They all look at it, pull it apart, so to speak. Find out the features.

As for white goods, as another manager told Ken, introducing his new sales assistant to the store, 'that side the birds tend to look after. Well, they *know* more about washing machines, don't they.'

People are seen as having gendered enthusiasms and transmitting these to customers. And certainly this is not mere prejudice. For example Fred, the Bunnett's branch manager, told us of a woman sales assistant who

> was a really good cook in her own right. And she actually bought a microwave while she was here, and took on microwave cooking and did it very well. She actually used to double up and do demonstrations for us. She would prepare things at home, bring them in and lay them out and it was really magnificent.

This phenomenon is gendered in its lack of symmetry. It is unlikely, in the gender circumstances prevailing in Bunnett's and the wider world, that a man would do this, or conversely that a woman might demonstrate, say, her home-made camcorder productions to customers.

The case of this woman microwave enthusiast is not conclusive of gender difference – no example ever is. It is however *symptomatic*. It is to such symptoms that those who are in a position to appoint staff respond. Reading similar cultural messages, interpreting them in the light of their own prejudices, they associate men with brown goods, women with white. They also associate women with clerical work and men with lifting work in the shops. Thus the Wonderworld manager, Harry, said of women: 'I think they're far more capable than what a man is of dealing with the problems of messing around with bits of paper and pens and stationery'. And managers are reluctant to allow women to join the clique of fellows that deal with the 'getting out' and on to display of products from the stockroom. Ken pointed out that in the stores in which he had worked, the lifting was a continual arena of male competitiveness.

> Most of the men who work at Wonderworld will attempt to lift heavy units down by themselves. This is in spite of being told not to do so and the repeated warning that it is in contravention of the Health and Safety at Work Act. . . I always asked for help even though it may only have been a fridge I wanted down. This usually led to my manhood being questioned. It's easy to see why very quickly younger males employed in such work fall into line.

Ken also pointed out that the gendering of this task deprived women of technical knowledge, since 'product knowledge is usually picked up through getting the unit out and running through the basics while displaying it'.

This technological gender hierarchy is the commonplace situation found throughout the hundreds of Arrow, Bunnett's and Wonderworld stores. It seems natural enough to most managers at branch level, but it is increasingly a source of annoyance at head office where strategic perceptions of commercial interests in today's world have begun to throw doubt on its commercial logic. We shall see that the training personnel are directing their efforts to breaking down what are now seen, from this elevation, as outdated stereotypes. Meantime, however, 'no matter what we try to do to avoid it', complained a training manager in a clipped and resigned tone of voice, 'it just naturally seems to happen: the women will stay with the white goods and the men will stay with the brown goods'.

In the light of this dichotomy between brown and white goods, manly and family goods, Arrow and Bunnett's environments, it is interesting to note that microwave ovens, until around 1989, appeared to have broken out of this frame. Shiny white though they were, they were sold by Home-Tec not only in Bunnett's, but also in Arrow, where they were a lone white good. Remember, they had been a lone white good in manufacturing and Electro too. What is it about microwaves? Somehow, in the early days, as one Home-Tec manager put it, the microwave oven 'wasn't necessarily a cooker, even though its function was to cook'. It was at first seen as pure gizmo, technology-as-such in the Japanese mode. As Home-Tec's microwave buyer put it, the microwave then

> was high technology, and high technology tends to be a male-dominated area. And rightly or not, the woman tends to abdicate that responsibility to the man. She says 'I don't understand it. Will you explain it to me?' I think that's definitely changing now. . . People see microwaves everywhere now. They go to a holiday chalet and that has one. People are much more familiar with them. They're part of the way we live. So the decision-making process [in purchase] is much like that for the toaster, the iron, the electric kettle or the washing machine.

In 1989 Home-Tec ceased selling microwave ovens in Arrow shops. It was no longer seen as value per square foot of floor space. Its ranking had slipped and other new gizmos had a better claim to eminence in Arrow's technological displays. In particular, camcorders had become the rage. The market for this product topped £275 million in 1990, with a year-on-year growth of 25 per cent. How could the microwave compete? Besides, with age, the essentially domestic nature of the microwave was ineluctably creeping up on it. 'In a way, perhaps microwaves were a misfit in Arrow,' mused Cecilia, Home-Tec's business planning analyst. She explained that microwaves had once been 'revolutionary', 'it was electronic, it was gadgetry, it was a box with flashing lights'.

> That's why you saw them in Arrow. After a while they became a standard must-have purchase. We moved them over to Bunnett's because . . . people understand what it is, that *it's just a basic kitchen appliance.* (Our italics)

Matthew, the advertising executive quoted in the previous chapter, threw an interesting light on the continually shifting social identity of technological consumer artifacts. Camcorders, he said, had arrived as indubitably masculine, a quality they had inherited in their descent from cameras on the one hand and videos on the other. Microwave today had become a female product, in that, unlike the early experimental moment, now 'most men don't want to get involved in using the product, in making decisions about it, being involved in food preparation in any way'. Music systems had also experienced metamorphoses in social identity, he explained. The development from the 'radiogram' to the small, inexpensive portable transistor had lifted listening from the 'family' to the individual (ungendered or mildly masculine) sphere. Then in the 1960s high fidelity technology arrived and the people 'who wanted to be involved in it or felt they could become part of it' were men. While trannies could conceivably be for grannies, the whole 'ownership' of hi-fi technology came to be seen as youthful and masculine. The group of young male cognoscenti, said Matthew,

> has this supposed expertise, make judgments about hi-fi. It's about the quality. It's perceived as something very expensive and people wanted authorities on it to help them out. People went to people who knew. You took advice, you bought magazines.

In its turn, with the advent of mini- and midi-systems, hi-fi subsequently slipped a little from its masculine perch. Nowhere near as far, however, as microwave has tumbled: into the kitchen.

Gendering the Customer

The contribution of the retail firm, then, to the life-cycle of the microwave oven, is not merely to deliver the microwave to the end user but to domesticate it, familialize it, 'whiten' it, demystify it and shape it into a socially wanted and accessible object. It does this by locating it in the appropriate category of goods in the store, projecting an image with it through promotional material, and 'dressing' a sales force in the appropriate gender mode. The retail actors however do not stop at designing the product/process and its sellers. They also, in a sense, design a potential purchaser.

We saw with the staffing of the shop that there is an uneasy correspondence (significant but incomplete) between the material sexual division of labour – where men and women are actually located with regard to technology and technical knowledge – and what is 'made of it', represented about it, as people talk and relate to each other and us. Similarly there is an elusive correspondence between, on the one hand, who microwave customers really are, how they really behave and, on the other, what people purvey about them. The meanings made from the retailer's experience of the shopper both reflect practices and shape them. There is an unclear dividing line between accurately *representing* the customer, *constructing* the customer and *controlling* the customer. The retailer of microwave ovens needs to construct a popular image of the shopper with some basis in fact in order to be able to predict appropriate sales techniques and promotional materials. At the same time the retailer needs to control the customer – both to clinch a sale and to lead the user into safe and responsible microwaving practices so as to avoid safety scandals that reflect badly on microwave manufacturers and retailers alike. In representation, construction and control of the user, concepts of

technology and gender, and the relationship between the two, are developed and deployed.

Let us look first at the image of the microwave buyer and user the retail actors construct, drawing on observation, hearsay and expectations. First, the trade materially distinguishes different kinds of shoppers and shopping patterns through the design and siting of its stores. A growing proportion of Arrow, Bunnett's and Wonderworld stores are edge-of-town superstores. Electric consumer durables are greedy for space and their sale benefits from the availability of car-parking. The superstore takes advantage of the lower land prices away from the city centre. In contrast to high street shops, where the peak shopping time is lunch-hour and the customer is seen as variously male or female and on foot, the busy time for the superstores is evening (they often open late) and weekends. The customer is represented as more likely to include couples, to be somewhat 'up-market' and to come by car.

Regardless of the kind of store, we found that microwave customers were seen very much as couples. Tracy, sales assistant in Bunnett's, said: 'I always find, myself, it's couples. Women will buy washing machines

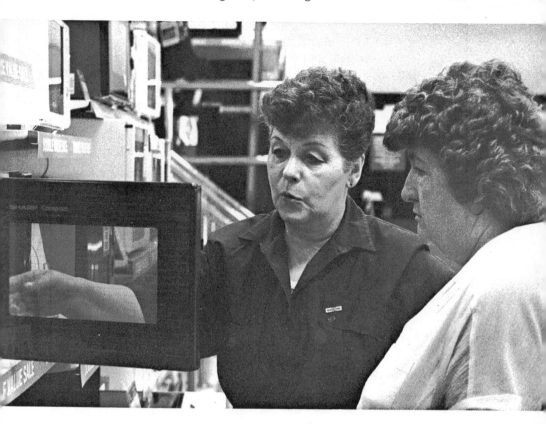

on their own, fridge-freezers. But cookers, microwaves, for some reason, I find, they always want their husband there. I don't know why.' Ken, reviewing the microwave transactions he had observed or in which he had participated over ten weeks at Wonderworld, remarked that 20 out of 24 had involved people of both sexes. Several people surmised that the prevalence of couples reflected the fact that a microwave is still in the main seen as a carefully considered purchase. As a non-essential with not insignificant financial implications it required negotiation between husband and wife.

It was also universally observed that the partners played gendered roles in the purchase. Ken reported: 'In most cases it was the woman who took the most active role in the proceedings. It was she who would be concerned about such things as colour, size and power levels. The man would hover around the area picking up on the odd piece of informa-tion.' The women was seen as being, of the pair, the one who would do the cooking. 'I think not many men, well, they know how to operate it, but most ladies do the necessary work of, carry out the cooking in a microwave,' said Harry, the Wonderworld branch manager. What was being portrayed therefore was a gendered division of interest, in which the woman would typically identify the need for a microwave, the specific features needed, and the aesthetics of colour and style. The man for his part would authorize the expenditure, take an interest in the technology, particularly the wattage of various models, and perhaps lend moral support to the purchase intention. Retail personnel said variously:

> She says, 'Shall we have that then?', he says, 'Yes'.

> He'll make the decision but she'll make him make the decision.

> The questions are asked by the woman, the decision is made by the man.

> I think the women in the household will drive the purchase in terms of 'I want one' and I think the men will come into it because of the technology associated with it.

Men were portrayed as being the ones most likely to have the responsibility and foresight to pay the extra to obtain an extended warranty: 'She thinks, "Well, I could have a new dress with that".' Ken reported that

> the final act in purchasing a microwave is the actual handing over of the money. Very often it now became apparent just why the man had even come to the store. This was because he was going to actually pay for the goods. 'So that's the one you want is it?', would in some cases be followed by a knowing male look at me from the husband. . . I have often been offered marital advice such as, 'Don't ever get married, son!' or, 'They're all the same!'

The trend today towards the purchase of more expensive models of microwave, particularly combination ovens, often as replacement purchases, may have increased the tendency to purchase by partners (or shopping in packs of two, as Ken put it). Men therefore feature in the retailer's thinking about electrical goods customers in two guises. They occur as independent male individuals in brown goods shopping, and as family members, husbands, heads of household, in white goods shopping. Women have a more unitary persona in this thinking: they are only domestic. The two masculine personae thus projected are superimposed in a widespread belief that may derive from the early days of microwave. It is that when men buy microwaves alone they are often buying them as presents for partners, and that (lured by their technological features) they characteristically buy complex, highly automated and high-powered models that their womenfolk do not like, understand or want – indeed that they are intimidated by.

In some accounts gender combines with age and class. (Ethnic group was never cited as a relevant variable.) It is the skilled manual working class, represented as 'earning hard and spending hard', prone to giving expensive Christmas and birthday presents, who are reported by retailers as having been the early microwave enthusiasts. The gift-giving male partner is paradigmatically the male head of a working-class family. Middle-class couples, especially men, are represented as being the more intellectually 'difficult' and discriminating purchasers. Perusing their consumer guides, they can be a headache to the sales assistant. Tracy said: 'They've got these bloody *Which?* magazines, and you think "Oh, my God! Disappear!".'

Older people are seen as the more recent, and more tentative, microwave converts. The older woman is represented as being the person most likely to be intimidated by modern electronic touch-pad controls with liquid crystal display ('macho because of its technology') and to prefer simple manual dials. A reassuring compromise is seen as the type of model that employs electronics, but has 'individual buttons which give a nice little beep when you press them, responsive things, interactive'.

These representations of the customer, expressing social reality while simultaneously constructing it, match those that describe and shape the seller, as we saw in the preceding section. They differentiate masculine and feminine, they associate both technology (as engineering) and authority (in this case 'the power of the purse') with the masculine, and cooking with the feminine. There is a certain difference, however, between the gender relations we have observed in the manufacturing case study and this study in retail. It is a difference of degree. For reasons on which we speculate toward the end of this chapter, in retail women have been permitted greater access to middle management positions. The sphere perceived as feminine is not so relatively diminished and devalued.

Controlling the User: Safety and Hazard

Constructing the microwave *purchase and purchasers* the retail actors build on, without necessarily expressing accurately, what occurs in practice. But if their perceptions are wildly far from the mark their selling will be less effective. It is equally critical that retailers build for themselves a realistic picture of *microwave use and user* and set up adequate communication with the latter, since their business reputation depends greatly on establishing the normality, popularity and above all the safety of the microwave cooking process. We saw in Chapter 3 that the manufacturers must make efforts to enrol the microwave purchaser as intelligent, responsible user. But although their sales consultants encounter a small percentage of customers, manufacturers do not, as the retailers do, meet every customer face to face. It is the retailer who mediates the translation of the microwave from factory product first to white good on the shelf, then to domestic appliance in the kitchen. The retailer is in the front line of customer complaints. It is the retailer who more often than not undertakes or subcontracts after-sales service on the artifacts sold.

The microwave cooking project has met with more impediments between conception and use than most other technological innovations in the electrical consumer field, due to a series of popular panics concerning its safety. Safety-in-use depends on both a reliable machine and a sensible user. When safety fails, each 'side' blames the other. We can select three issues the retailer has had to handle in a customer-control strategy. The first is risk of technical failure. The second is danger from food poisoning caused by incomplete cooking. The third is radiation hazard. In retail actors' narrations of these dangers, and what they see as their customers' responsibilities, it is possible to detect the shaping of a social identity for the new technology and a simultaneous reshaping of feminine gender identity to fit it for the challenge this new technology represents.

Technical Failure

The retailer is caught in a difficult situation. There is sound commercial advantage, as well as good consumer sense, in facing up to the fallibility both of the microwave oven and its user. Breakdown does occur, misuse is common, sometimes parts are faulty. Efficient after-sales service is attractive to the shopper. Yet too much emphasis on the frequency of breakdown and need for repair frightens the customer away. Public confidence in consumer durables depends upon them proving durable in fact. The contradiction for the retailer is intensified by the fact that the sale of extended warranties (which as we saw in Chapter 1 is currently a significant source of profit for electrical goods retailers) must inevitably involve the selling of mistrust in the technology. To sell this insurance against breakdown, as one store manager deplored, 'I have to sow doubts in their minds'.

   The way retailers attempt to wriggle out of the contradiction is, first, by distancing themselves from the manufacturer. We asked Home-Tec's advertising agent: 'Isn't there a complication for you in selling warranties if you're at the same time trying to advertise reliability?' He answered, 'Yes, but you see, we're *not*. That's the manufacturer's job. The manufacturer is trying to advertise reliability, absolutely. And the retailers are not in that game.'

   Secondly, the retailer invokes the very modernity of the technology to suggest to the prospective buyer that the satisfactory survival of the artifact is beyond the owner's control and influence. The way electronic equipment is described emphasizes that it comes 'all of a piece' and consequently if it breaks down, cannot be repaired piecemeal. The opacity of the high-tech artifact is emphasized to persuade the owner that he (or she) is incapable of intervention in it. Indeed, breakdown is ascribed more to fate than anything else. The purchase of a warranty becomes a game of chance. This again, however, runs counter to the retailer's interest in instilling into customers the idea, first, that the technology is not difficult to understand, and secondly, that they must be responsible users, follow instructions to the letter if they themselves are not to be the source of breakdown.

   Like other electrical retail multiples, Home-Tec is obliged to address the

problem of technical failure of the products it sells. It organizes financial services to sell extended warranties to buyers of appliances, usually for two or four years beyond the manufacturer's one-year guarantee. Other organizations are here enlisted peripherally in the microwave-world as insurers or underwriters. Sales assistants have been intensively trained in methods of selling this insurance. In a psychologically difficult ploy the salesperson must confront the anxiety the customer is feeling in writing an alarmingly big cheque, or committing herself or himself to future outgoings to pay off a credit purchase, in order to urge yet further expenditure on the basis of 'a little more now to avoid much more later'. 'It's peace of mind. We sell them on the idea that you'll never have to put your hand in your pocket for any repairs.' A particularly telling argument, it seems, is the idea that breakdown just before the end of the warranty period may well result in the manufacturer replacing the already well-used machine with a brand new one.

The second organizational response by Home-Tec and other such firms to technical fallibility has been to undertake its own commodity scrutiny and testing. Home-Tec subcontracts testing to Tec-Care plc, which it commissions to assess all the commodities it considers stocking. Tec-Care checks the artifact for safety: does it comply with national safety standards? And performance: does it do what it's supposed to do? Finally it asks: is it sensibly serviceable? how can it be disassembled? will parts be obtainable? Tec-Care also carries out after-sales service of brown goods sold through Home-Tec's chains, Arrow and Bunnett's. Combining testing with service helps the company present a responsible image. Being more 'up-front' about technical fallibility is a recent development. A manager at Tec-Care said: 'Home-Tec didn't like to admit that Tec-Care existed. . . They didn't want [people] to see that a machine may go wrong. But now the customer expects, demands, that the machine is supported.'

Likewise, Home-Tec subcontracts to another specialist company (which we called Nationwide After-Care plc) the after-sales servicing of white goods, including microwave ovens sold through Bunnett's. The financial liability for the cost of repairs carried out by After-Care falls variously on the manufacturer, the insurer or the user. Here again, the retailer is in something of a cleft stick. The company is frequently torn between wishing to hold the customer responsible for misuse of the artifact and wishing to maintain a reputation for hassle-free service. Nationwide After-Care plc was at pains to duck the brickbats. If it felt the customer was to blame

> what we would do is say to the customer, 'I'm sorry, the manufacturer won't undertake this under warranty. It'll cost you so much to have this put right.' Invariably, the customer will then go back to the retailer she purchased it from, there'll be a big row, and they'll come to some agreement. Or the manufacturer will say, 'We'll do this as a one-off.' If *we* were to take it up [with the manufacturer] they'd probably turn round and say, 'No, we're not paying you for this, it's customer misuse.'

In tune with the new openness, Home-Tec is permitting Tec-Care and Nationwide After-Care to emerge from the shadows and become personable. Bunnett's bigger superstores are now opening 'repair desks' seven days a week. The idea is that people can now 'take their machine to a *man*'. (They mean 'a person'. Being a repair specialist, it is likely to be a man.)

The retailer's dilemma over technical fallibility reveals itself in discussions of fire hazard. The press and other media often carry reports of fires occurring in microwave ovens. Usually the food that ignites is a sweet and fatty item such as a Christmas pudding, cooked for too long. Both the sales staff and service engineers are outspoken in their condemnation of the irresponsible user. 'Imagine if she leaves a little bit of food around the door or some of the other places which in the book it tells you to clean after every use, you can get all sorts of interesting faults which can write off ovens,' said a Tec-Care manager. A store manager complained: 'You've never heard any outrageous tales about putting a cat in a washing machine to wash it. But you hear lots of outrageous tales about putting a cat in the microwave to dry it!'

Food Poisoning

The many and colourful insults levelled at the users behind their backs contrast sharply with the respectful relationship sought on the sales floor. The construction of the user as irresponsible, however, contributes to the importance retailers ascribe to instruction and instruction materials in effecting the translation of the microwave from commodity to domestic equipment in household use. The adequacy of instruction was widely questioned during the listeria food poisoning scare of 1989. The events have already been described in Chapter 1. Suffice it here to recall that cases of food poisoning from inadequately heated and sterilized food, particularly listeria poisoning from bacteria remaining active in cook-chill recipe dishes of the kind commonly heated in microwave ovens, were reaching the media. As we have seen, the Ministry of Agriculture, Fisheries and Food carried out and published the results of tests of various models of microwave. The findings revealed some models to have a fault in the wave distribution that was permitting 'cold spots' to persist in heated dishes. Although some manufacturers were eventually named and blamed for what were clearly design deficiencies, some of the responsibility was passed back to the user who, it was insisted, must be the final arbiter of whether the food is safely 'done'. New regulations to standardize ovens' power ratings and to classify food dishes in 'bands' according to time needed to cook them at given power settings went some way towards reassuring public opinion.

Retailers were ready to question their own practices too. They set up enquiry lines and offered refunds on models that had failed the tests. In terms of written instruction, they felt that, working with the manufacturers,

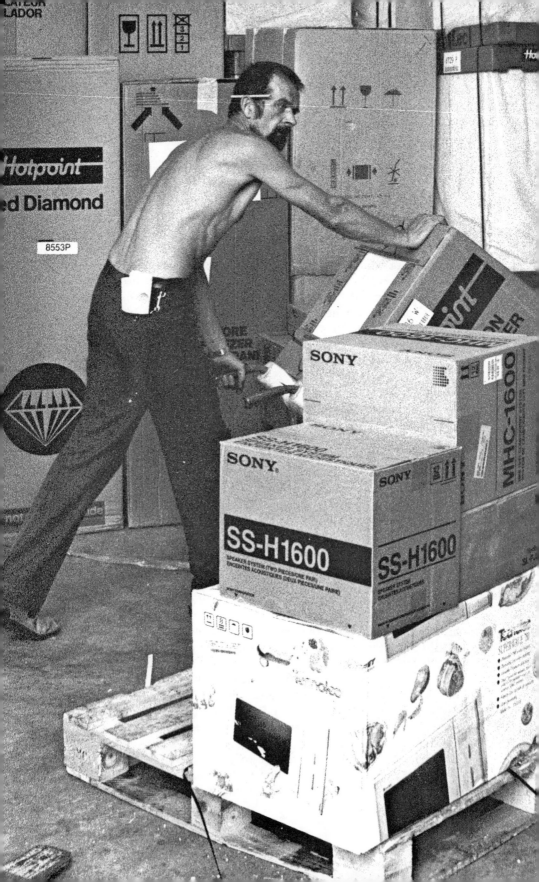

they had at different times in the past tried everything. They had produced information the user was meant to read and throw away, such as peel-off stickers. They had produced material the user simply could *not* lose however hard they tried: flip-page charts and manuals ingeniously slotted underneath the oven, or attached to its feet. They had tried covering books in hard plastic sheeting to make them more durable. They had favoured more explicit icons impressed into the microwave ovens' control panels and doors. They had asked manufacturers both for more pithy instructions and for more wordy versions. They had tried simple flow-charts and more interesting and detailed cookery guides.

Privately they despaired of microwave users of whom it was said they 'have a limited attention span', 'can't follow instructions', and 'are too impatient'. The user was blamed simultaneously for failing to respond to the new demands of a technological innovation and failing to use the common sense born of traditional cooking. Even the Consumers' Association turned its attention now to whether the public was intelligent enough to understand either existing or proposed instructions.

Who exactly was this incompetent 'user' the retailers had in mind? Although there was some attempt to use sex-neutral terminology, in the course of conversation it was impossible to avoid for long using the personal pronoun. What slipped out was usually 'she'. If fire and breakdown issues could on occasion be dealt with in genderless language, in the matter of food poisoning it was more difficult. We shall see in the following chapter how responsibility for nutrition and health is still firmly located in the popular mind with the woman, wife and mother.

Radiation Hazard

The earliest scare concerning microwaves, in the late 1970s, just as they were beginning to gain popularity, had highlighted the risk of irradiation. Microwave radiation has been blamed for various maladies, including cataracts. Fear of microwave energy is related to a more general anxiety about the effects on modern communities of living in close proximity to increasing amounts of electromagnetic equipment. It also has resonance for those who believe that some kind of deliberate harm to groups of people seen as political enemies has at certain times been attempted through 'zapping' by means of unspecified kinds of waves or rays.

Lay opinion is divided concerning the validity of such anxieties. Technical researchers, however, claim that no investigation of microwave ovens has yet revealed any danger of this kind. Of course radiation at the frequency of microwave emissions can cause heating and burning of the flesh, just as it can heat and burn food. To burn a user, microwave energy would need to escape from faultily constructed ovens or badly fitting doors. The researcher at the Consumers' Association responsible for test-ing and reporting on microwave ovens told us that instances of such leaky ovens were in his experience rare. No hard evidence has yet been produced of other less visible or less immediately harmful radiation effects. Microwave specialists, from their committed viewpoint, told us unambiguously: 'There really is no problem'; 'It's laughable'; 'I would say now that there's not a proven case in the world that a microwave cooker has caused a health hazard in that sense.'

Retail sales assistants and store managers nonetheless often have to answer questions from potential buyers concerning radiation hazard. They are trained to answer patiently and reassuringly. To us, however, they sometimes expressed exasperation. The Home-Tec customer services manager was incredulous that a woman could have phoned the company with the worry that the microwave she had bought in a Bunnett's store might have been the cause of her giving birth to a deformed stillborn baby. It is particularly women who are represented as fearful of radiation. We found the term 'technofear' in common use to describe variously an alienation from technology; a fear of its unknown powers; and a nervousness concerning one's ability to understand and use it. It was widely supposed that technofear was a female characteristic.

> We do feel, perhaps for a woman for instance, that there is some slight technological *fear*.
>
> I suppose my inbuilt bias would be to say that men have less technofear than women. Because they're into the camcorders and the hi-fi and all this new technology, you know. They're obviously less likely to be daunted by something like a microwave.
>
> I'd think it's more likely to be the ladies.

One woman, the home economist Marion, explained the concept at greater length.

> Technofear is about the woman's conception of herself being unable to control technology, because [the microwave is] a new cooking method, not like her hob or oven. . . I think there is still a large group of women who, for some reason, whether it was through something that happened when they were children, or at some stage in their lives, have become more resistant to technology. If you hand them a calculator they're more likely to be worried about suddenly being confronted with it. Or with the panel of a microwave oven. They'll think 'jeepers, creepers!'

The woman user is often criticized for believing the microwave oven to be a 'magical machine' or a 'magic box'. By this they perhaps mean something similar to the 'black box' dear to the sociologist of technology: a device of which one sees the inputs and the outputs without understanding anything of the internal mechanism. We ourselves did not notice any marked difference between lay women and lay men when it came to explaining the way microwave ovens 'worked'. Many people say, 'I know it's supposed to cook from the inside out'. The specialists laugh at this notion and use it as evidence that most people do not understand microwaving. They say that, quite the contrary, the microwaves penetrate food only to a depth of two or three centimetres. Yet, correct as their version is, in a sense the popular view is also perfectly accurate. No external heat bears on the food in microwaving. It is the vibrating molecules themselves that generate internal heat and 'cook' the food.

We heard one or two dissenting voices in the retail trade that spoke up for women's technological competence. One home economist pointed out that women are less inhibited than men in saying they don't understand a given artifact. Men often know as little as women, but they bluff their way through because a man is more ashamed to admit to technological ignorance. A white goods buyer (male) for Wonderworld protested against all the men he continually heard running down women's technological competence, designers and product planners who argued for simpler microwaves with more old-fashioned manual dials, holding this or that innovation to be 'too complicated for women'.

> Let's face it, it's women that use these appliances, as I've pointed out time and again. . . Women are the people that.use in the home the high-tech products. It's my wife that uses the video recorder, not me. And my wife uses the touch-control and auto-sensor microwave, not me. . . I think if you took the full touch-control model to the marketplace, it would be everything a woman wants. A plush, wonderful membrane that's got pre-set programmes. . . it's got to be what she wants.

When Sexism is Counter-Commercial

We saw in Electro UK (as reported in Chapter 2) signs that at least one progressive and far-sighted manager found the commonplace stereotyping of women as non-technological and unsuitable for management jobs to be something of an embarrassment. This anti-sexist turn of mind was evident too among some individuals in the retail trade. The Wonderworld buyer quoted immediately above was one of these. While he, and the manager who championed Carol in Electro UK, can be represented merely as isolated liberals, it is also true that they give expression to an alternative mode of thinking slowly emerging within contemporary business organizations. Both Home-Tec and Wonderworld, for instance, had formal, published, sex equality policies. Pressed in particular by more senior women, there was a commitment in both companies to correcting the imbalance between numbers of women and men in the upper grades.

In both these retail chains we found that training programmes for sales staff had recently been set up that reflected the anti-sexist policy in interesting ways. Management had been needled by adverse comment in a *Which?* survey on the standard of shop-floor service and advice in the larger multiples. They were worried, said a Home-Tec's manager at head office, that customers were saying, 'Huh! bloody Home-Tec, you know I went into a store the other day and this idiot served me, and he didn't know what he was talking about.' They were beginning to understand that 'the days of stacking 'em high and selling 'em cheap – I won't be here tomorrow' had gone. Both Home-Tec and Wonderworld now had training departments that produced booklets and videos for distribution to their areas and branches. Store managers were conducting training sessions, usually first thing in the morning, on one or two days a week. Competitive tests were being held, with prizes for those sales assistants who scored high. There were two main topics within such training programmes: *selling technique* and *product knowledge*.

The emphasis on *selling technique* reflects the assertion that 'selling's a profession', 'a salesperson's made not born'. Broadly speaking it involved instruction in how to find out, by sensitive questioning, what the customer has in mind; how to assess her or his needs; how to describe the relevant product range in terms of features, advantages and benefits; how to offer credit terms; and how to clinch a sale, following up with a persuasive argument for an extended warranty. There is a strong emphasis on psychology. As sales assistant Tracy put it, 'people don't wanna open up to you, so you gotta find ways to open 'em up'. There was also a strong motivational component to the training, something of which comes across in the following exchange between ourselves, a bemused Tracy and an older male colleague, Rick.

Tracy: It's all to do with 'assignment excellence', because we've all got a different outlook on it now.
Interviewer: What does it mean, 'assignment excellence'?
Tracy: 'It's Up To Me, It's Got To Be Me.'
Interviewer: Is this sales target chart on the wall part of this?
Tracy: Yes, anything you find in here. It's all to do with *you*. It's very difficult to explain it. It's – [*stops*].

Rick: [*helping her out*] It's winners.
Tracy: It's what you're here for, innit.
Rick: It's winners, she's a winner.
Tracy: Yeah.
Rick: It's winners and it's losers. That's how it is. But if you lose, you come back on Monday.

The trainers have a problem concerning the most productive approach to gender in the customer. Stereotypes of masculine and feminine domestic roles are not an accurate reflection of reality. On the other hand, there really are strongly marked patterns of relationship between women and men, and teaching a perceptive approach to shoppers must involve a certain 'knowingness' about conventional gender relationships combined with sensitivity to individual feelings. The complexity of this was expressed by Tracy:

You can't go up to a customer and think 'yeah, she cooks all the meals' and talk to her direct. Because you've only got to repeat yourself to the bloke, or otherwise he doesn't like it because he's being kept out. And he's the one who's paying for it, he's the breadwinner. . . If a couple come in, you've got to talk to both of them.

Without gender sensitivity, a salesperson would come across as naive or unconventional, and would fail to generate the needed trust. Yet increasingly management appear to consider it advisable to de-gender the customer that is visualized during training. The formal training emphasis, we found, was on finding out, not permitting stereotypes to predict lifestyles and needs. Both Home-Tec's and Wonderworld's training programmes attempted a de-stereotyping of sex roles. In the training videos a man sells a man a microwave. A woman sells a man a washing machine. It is interesting that the cover of the accompanying booklet, instead of using drawings identifiable as women or men, used a totally unisex plasticine 'mannikin'. Besides, the introductory video to one of the training programmes employed professional actors in a comic sketch of caveman and cavewoman cooking a pig's head in a pot over an open fire, washing the baby's clothes in a stone tub and sweeping with a broom. The man was the main butt of the humour, because he was incompetent and demanding while his missus did all the work. The spoken message was: 'How much easier their lives would have been if

they could just have popped down to [our store] to have all of their domestic needs taken care of in a thoroughly modern way.' An additional, unvoiced, message was that not only the technology but also these sex-stereotyped behaviours are out of date.

Teaching *product knowledge* also calls for continually new thinking. It is a serious educational challenge, when a store may well bring into stock a total of 1500 new models every year, involving a total of 15,000 features. Usually the products are grouped generically and made the subject of video tapes, one for example dealing with camcorders, another with microwaves. The various product modules are sometimes sponsored by manufacturers of the artifacts concerned. Home-Tec was giving equal importance in product knowledge training to white and brown goods. In Wonderworld we found more training modules dealing with brown goods, which were perceived as being more the complex lines and as constituting more of a sales challenge.

While there was no gender distinction in the training for sales technique, there was an asymmetry apparent in product knowledge training for brown goods and white goods in both companies. Training departments aimed for product knowledge for every sales assistant across the board. The trainers clearly perceived two distinct challenges here: to persuade men that white goods are worthy of their attention; and to persuade women that the technicalities of brown goods need not intimidate them. The first involves overcoming male techno-snobbery, the second involves overcoming female technofear.

The training officer at Home-Tec said that despite the increasingly advanced level of white goods technology – 'you've got these super-chill refrigerators, ecological washing machines, microchip technology' – men still considered this side of the business beneath their dignity. 'We're saying that everybody in Bunnett's, even 17-year-old Jimmy who loves audio hi-fi, has got to be learning about white goods.' By the same count, 'Someone that's stuck with white goods and won't go on to brown goods, is, if you like, forced to address this.'

Like Home-Tec, Wonderworld appeared to have adopted a strategy of breaking down sex-segregation in their stores. One Wonderworld training video employed the device of a young saleswoman specifically asking the training officer for help in extending her product knowledge to brown goods. Although the explicit message in both Home-Tec and Wonderworld training is that 'sex differences are minimal and should be discounted', the film-makers find it difficult to avoid reinforcing sex stereotypes in all manner of subliminal messages. An example of this will be further discussed in Chapter 6. A further instance of the contradictory thinking (again from Wonderworld) is that on one training course reported by Ken the prizes for top performers took the form of men's ties for male winners, and 'an exciting range of women's lingerie' for female winners. The occasion as described by Ken involved a male trainer accompanied by a

> very 'pretty' woman who sat by his side and occasionally handed him his props, which seemed to be the whole purpose of her employment. . . He then proceeded to pass the knickers etc. around the assorted audience, asking in general tones, 'A bit of all right, isn't it girls!'

We have noted nonetheless that an organizational strategy to counter sex-stereotyping was found in both retail chains. Both spoke of themselves as 'Equal Opportunities employers'. Even at Electro they had begun to sense some advantage in breaking down certain aspects of sex-segregation in employment – or perhaps simply to sense disadvantage in resisting it. We shall return to these developments in the final chapter. Meanwhile we move on to the third important moment in the microwave's life-cycle: the moment of use.

5

Cooking and Zapping

A microwave oven may be purchased for one of many reasons – as a present, because it 'seems to be the thing', to solve a perceived cooking/eating problem. It is (as the sales staff see it) characteristically bought by a couple, sometimes by a man for a woman, but also quite often, we may be sure, by a woman or a man on her or his own account. Microwave ovens are bought casually, hopefully, dubiously, discerningly. One way or another they find their way to a kitchen where they compete for space on the worktop with the electric toaster, the electric kettle, the electric liquidizer.

The microwave oven arrives as a *social* entity. We have seen how it originates in the very human processes of design and production, and how its distribution and sale also involve complex social interaction. Now it brings its persona into a new set of relations – those of the household. This chain of sociality linking the parts of the microwave-world can with reason be called 'technology relations'. However, as we have seen, what is involved here is simultaneously 'gender relations'. Even the microwave as artifact has an interestingly gendered meaning. It has indeed manifested some 'interpretive flexibility', changing its gender in the process of its translation from product, to commodity, to appliance. It devolved from the thought processes of the (male) engineer, passed through that early moment in which it was celebrated as state-of-the-art technology for which men were seen as likely buyers, and became a mere price-competitive (family) white good. Now we find it, an unremarkable appliance, in the workaday feminine environment of the domestic kitchen.

A straightforward technology impact study might now ask, 'What effect does the microwave oven have on cooking, eating and other aspects of domestic life?' Our perspective, however, leads us to doubt that we would find any simple one-directional bearing of the technology on social relations. Rather we would suppose that an interaction will now take place. The artifact will play a part in shaping new social relations along with other concurrent developments outside, but impinging on, the microwave-world. We can also expect the circumstances of the microwave's kitchen use to modify future models. Many possible directions for the microwave have, it is true, now been passed by. Closure, of a kind, has occurred to enable mass production to take place. Yet what microwaving is to become is not quite fixed.

What matters to the microwave's originators and distributors is whether it is put to the use for which it is intended, whether it truly 'catches on' as the instrument of a new cooking/eating practice. And here the picture is complicated by the fact that manufacturers produce not one microwave model but a *range* of models, with rather different intentions for each. They cover their bets by producing and selling everything from a cheap and cheerful box for zapping instant meals to a complex culinary technology. The latter, as we saw in Chapter 3, itself embodies two contradictory trajectories, one towards combined heat sources for the experienced cook, the other towards total automaticity for – whom? The manufacturers themselves are not sure. The range is continually being modified, adapted, extended in the light of sales performance. (So too is the construction of the putative user.) Some features (of both) will continue and develop, others will be dropped. Through the power of the purse the end user has a modest influence in shaping both herself/himself and the technology.

Beyond the decision to buy or not to buy, however, the user has one further option, albeit a self-defeating one: to buy and not to use, either the oven itself or some of the features that have been paid for. The chances are that where features remain unused they will not be sought in a second purchase. On the other hand, manufacturers may well see advantage in sustaining price by offering desired features only in combination with undesired features. There is a guaranteed match neither between purchase and use, nor between supply and demand.

The cooking process through which the user or boycotter of microwave ovens and their features (woman, man, couple, household) produces food is as much a gendered practice as the engineering that produced the oven itself. Gender relations, then, shape technology-in-use, and give it a certain meaning. Conversely, a new technology entering the home opens up a new arena in which, and new material on which, gender relations will act.

The Changing Experience of Domestic Life and Labour

Microwave cooking is no more than one recent instance within a continuum of technological change in domestic life. In the course of the Industrial Revolution, let us say from the mid-eighteenth century, many of the goods that had traditionally been produced in the household began to be manufactured for the market. As farming was rationalized, labourers lost their grazing rights and many moved to the towns, animals were less often reared and slaughtered at home. Instead, meat was processed on the conveyor belts of industrial abattoirs and purchased for cash. Beer was now brewed in breweries; soap, candles and cheese increasingly factory-made. The 'domestic' technologies that mediated these changes were outside the household but they nonetheless effected a revolution in household labour processes.

As the cities grew and the rural population was depleted, new infra-structure – brick-built and slate-roofed houses, piped water, coal supplies to replace wood for heating – reduced the overall amount of labour called for in household maintenance. Some tasks were added, however. Washable fabrics such as cotton and linen replaced felt and leather so that laundry became a time-consuming chore.

Change in technologies of industry and household was inextricably related to change in the social division of, and arrangements for, labour, both paid and unpaid. The Industrial Revolution brought into existence a large class of waged labourers. Many women, particularly unmarried working-class women, were drawn, like men, into industrial production. Others left their family homes to go into domestic service, for the new industrial bourgeoisie and some skilled manual workers employed lower-class women as domestic servants. Some of the time and energy of better-off wives and mothers was freed. At the same time the entrapment of lower-class women within the patriarchal home, their dependence on the male wage, was less complete. But women as a sex did not lose their responsibility for housework: working-class women did it, women of more affluent classes managed it.

Ruth Schwartz Cowan in her history of domestic technology traces the unequal effects on the sexes, in this class- and gender-structured context, of nineteenth-century technological developments. She concludes: 'industrialization served to eliminate the work that men (and children) had once been assigned to do, while at the same time leaving the work of women untouched or even augmented' (Cowan, 1989: 63). The need for whittling and leather work for instance was diminished by the growth of industrial joineries and shoe factories. Sewing was not so rapidly or completely industrialized, and as mentioned above, clothes-washing increased.

Beyond a certain point, the household instead of becoming emptied of more and more of its domestic tasks began to be equipped with machinery and appliances to achieve them in different ways. In the twen-tieth century, industrial expansion took the form less of provision of collective services than of production of consumer durables. Millions of individual households were seen as a valuable market by the manufac-turers of consumer goods. We did not move to eating our dinners in communal canteens; rather we cooked individual family meals on more and more efficient domestic cookers. The working class did not join the middle class in sending their clothes to laundries; instead more and more families obtained electricity-driven 'twin-tubs', and eventually elec-tronically controlled washers and tumble-driers. Children continue to be cared for more individually than collectively. Few goods today are delivered to the home; the supply of household provisions depends on repeated forays by household members to shops. The family has been pared down to its nuclear core and it has lost to the state some of its educational and health-care responsibilities. Yet the home remains the unchallenged centre of most people's lives and the site of unremitting work.

Technological innovation in the household, it seems, reduces drudgery, but not the *time* spent on housework. Research in several countries has recorded average housework hours. Ann Oakley (1974) reviewed some such studies carried out between 1929 and 1971, a period in which the mass production of consumer products, white goods especially, was a growing feature of British manufacturing. The conclusion appeared to be that the average time spent on housework had either stayed constant or increased slightly. A Swedish study comparing the 1980s with the 1930s confirmed this.

> Greater access to household technology and merchandise did not reduce married women's housework hours. The potential reduction in housework time was counter-balanced by higher overall consumption of goods and services. The result was the same or more housework time with no correlation to married women's labour force participation. (Nyberg, 1989)

In a multidisciplinary review of literature on the effect of changing utilities, appliances, foods and services Christine Bose and co-authors argue that, of itself, responsibility for the purchase, operation, running and maintenance of so much machinery has been a significant addition to household work (Bose et al., 1984).

In a later analysis of time-use surveys carried out in the 1970s and 1980s Jonathan Gershuny contested their finding that even in periods of diffusion of domestic technologies 'domestic work tends to remain constant or even to increase over time'. He re-worked some of the data from these surveys to show that over three decades from 1961 in the UK and the USA 'certain sorts of domestic work time have in fact decreased markedly' (Gershuny and Robinson, 1988: 539). As Judy Wajcman points out however, the reduction Gershuny uncovers is in only one of three facets of housework: 'routine' domestic chores. Two other important facets – childcare, and shopping with its related travel – became more not less time-consuming over this period (Wajcman, 1991).

The consensus seems to be that overall housework hours are remaining more or less constant but the composition of household work and responsibility is changing. The change is in the direction of ever-rising standards both in the provision of facilities and services and in the nurturing of relationships (Bereano et al., 1985). Home life in the late twentieth century in the industrialized Western world is quintessentially the guarantor of the 'quality of life'. Arnold and Burr (1985) contrast the effect of science and technology applied to capitalist production, where the upward spiral of product and service quality is determined and limited by cost and competition, with that of science and technology applied in the family, where there exist no such limits to 'progress'. Of course the manufacturers of appliances have stimulated demand. It is, however, not only the pressures of advertising but the beliefs current in the wider society, reflected and exploited in advertising, that says when the housewifery is 'good enough', establishes what is at any given moment an acceptable standard of nutrition, cleanliness and care. Expectations spiral upwards.

The equilibrium struck between housework time, technology and quality of output has been an effect of the interrelation of capitalist economics and class/gender relations in and beyond the (changing) patriarchal family. The key issue for a gender analysis is of course *who* does the housework. Traditionally it is uniquely women who take responsibility for the daily cycle of cleaning, cooking, washing and caring. Men are customarily, like children, those in a position to enjoy being provisioned. Men have normally been expected to supply a wage to support the household, and carry out certain less regular tasks such as repair and maintenance of building, vehicles and equipment.

The shift from a feudal rural society to an industrial urban economy, as noted above, removed from the home and located *in the factory* more of men's than women's traditional tasks. Given the observable cultural affinity between masculinity and machinery it might have been expected that, in the later period during which household tasks shifted from hand to machine *within the home*, men might have come to do more housework. Yet in industry, while men were (and are) the engineers with a controlling knowledge of technology, it was in fact women who became the operatives on the machines men produced and managed. Likewise in the household. Successive appliances came into the home as engineered products, but this was in no way incompatible with women continuing to be the labourers – operatives of domestic technology.

Opinions differ as to whether the domestic technological innovations of the second half of the twentieth century have coincided with a transfer of housework time from women to men, or whether it has had a contrary effect. Jonathan Gershuny finds, in the UK and the USA, 'that. . . women in the 1980s do substantially less housework than those in equivalent circumstances in the 1960s and that men do a little more than they did (although still much less than women)' (Gershuny and Robinson, 1988: 537). He is agnostic as to whether this effect has been caused by technological change or changing attitudes. Charles Thrall, however, concludes from a study of 99 families in the USA that time-saving technology is far from being a subversive force. On the contrary, by increasing productivity, it has enabled traditional family values and relationships to survive the entry of so many married women into the paid workforce. His study found that technology has a conservative effect on social relations. Indeed, 'when families have an item of equipment which is used for a particular task, they are likely to be more traditional in their division of labor for that task than are families that do not have the equipment' (Thrall, 1982: 194). Anne Murcott analyses the images of women presented in popular cookbooks and household manuals and concludes that not only do women gain new tasks as the old ones are 'lightened' but the very lightening of the domestic load has the effect of keeping women (as opposed to men) doing it. It has legitimized the expectation that a woman will continue to carry responsibility for both career and family (Murcott, 1983a). Domestic technology powers the flight of Superwoman.

Women's earnings have increased the household's income. Women's earnings and the appliances they buy, together with the continuing expenditure of women's time and energy in housework, have together made it possible to maintain or increase the quality of life for the rest of the family. Ruth Schwartz Cowan describes the resulting situation in the United States. The same could be said of the UK and many other countries.

> When, in the decades after the Second World War, our economy finally became capable of realizing the potential benefits of these technological systems, the individual household, the individual ownership of tools, and the allocation of housework to women had, almost literally, been cast in the stainless steel, the copper, and the aluminium out of which those systems were composed. . . A thirty-five-hour week (housework) added to a forty-hour week (paid employment) adds up to a working week that even sweatshops cannot match. With all her appliances and amenities, the status of being a 'working mother'. . . today is, as three eminent experts have suggested, virtually a guarantee of being overworked and perpetually exhausted. (Cowan, 1989: 212)

The conclusion of many researchers, then, is that the domestic gender relations into which new technology enters mould its use to their own shape.

Two studies of the use of microwave ovens suggest that microwaves have been no exception to the rule. A detailed study of the introduction of the microwave into 17 households in Sweden found that the share of food preparation and cooking between women and men remained unchanged. The only difference found was that children were more likely to heat up food (prepared by someone else). Women were overwhelmingly the cooks and there were no signs of increased cooperation between partners (Johansson, 1988). Among more than 300 British households surveyed by questionnaire in 1987 the finding was similar: 'Microwave cooker ownership does not encourage men to cook more often, although it seems that children are more likely to take part in household food preparation and cooking' (Burnett, 1990).

The gender relations that have thus been sustained through waves of technological change are relations of inequality. The inequality is less extreme, the male dominance less complete, in the twentieth-century family than it was in the formally patriarchal family of the nineteenth. Women have gained access to higher levels of education. They have increasingly earned their own wages in paid employment. More women head their own households or live alone. Nonetheless where women do live in partnership with men inequality continues in tangible ways. One important expression of inequality is the way heterosexual familial relations involve women serving the needs of men, children and other dependants with their labour in the home in a way in which they themselves are rarely served (Chabaud-Rychter et al., 1985).

The family-household system, as Michèle Barrett points out, provides a uniquely effective mechanism for securing societal continuity over a period of time. 'It has proved a stable (intractable) system both for the reproduction of labour power, and as an arrangement to contain personal life in the face of major social upheavals' (Barrett, 1980: 212). The housewife-mother and her housework remain a practical and symbolic lynchpin for structures of class and of male dominance. Her washing and cleaning keep family members healthy; they also maintain standards of order and propriety. The training she gives her children improves their chances of survival but also inculcates discipline and respect for authority. Of all the facets of housework, however, the provision of food is, materially and symbolically, central.

In a study of 200 women with young children carried out in 1983 Nickie Charles and Marion Kerr set out to understand the ways in which food practices contribute to the reproduction of the social order. Their sample came from a range of social class backgrounds. All but ten were married or cohabiting with men. Using repeated interviews and diaries they sought to find out how social divisions of gender, age and class are reproduced through food provision (Charles and Kerr, 1988). Despite changes in many other aspects of family life they found striking continuity and similarity in practices and beliefs concerning eating. Almost all the women in the study, they found, were even now the main providers of food to the family. Where men and children participated it was defined as 'help' for the woman. Men were given more food, in particular more high-status foods including meat, and their likes and dislikes were given priority. Women subordinated their own food preferences, and tended to 'treat' men, and to a lesser extent children, to their preferred foods. Food was used to mark class distinctions, celebrate occasions, generate pleasure, forge togetherness, instil hierarchy, win love. The way to a man's heart was clearly seen as being, as the old saying has it, through his stomach.

The women interviewed by Charles and Kerr placed great importance on the role of the woman-wife-mother in providing food that is 'wholesome' and 'nutritious', the epitome of 'good, home cooking'. Above all, women were striving for the regular provision of 'the proper meal', 'the family meal', which it was felt could only be cooked *by* a woman and could, essentially, be eaten only *as* a family: when the father was away, 'proper' meals were often dispensed with. Women on low income unable to feed their families 'properly' felt this to be a serious deprivation, not only in terms of nutrition, but as an indication of failure as a family. Food practices, these authors concluded, reflect class divisions and 'reproduce the patriarchal family, characterized by the authority of the father and the subordination of the mother and, at the same time, the authority of both over the children' (Charles and Kerr, 1988: 225).

If the provision by women of appropriate food in appropriate ways serves to maintain traditional hierarchical gender relations in the family, there have been many societal trends tending in a contrary direction. Families can scarcely behave in traditional ways without homes, and 170,000 households in Britain were homeless in 1990 – an increase of 15 per cent on the year before. Men's authority in the family is eroded by unemployment, and by the early 1990s 7 per cent of men in the UK were looking for a job. Marriages are less stable: 151,000 divorces were carried through in England and Wales in 1989 – the number had doubled since 1971. The proportion of households in Britain consisting of a single person increased from one in eight in 1961 to one in four in 1990. In 1990 28 per cent of all children born in the UK were born outside marriage. The proportion of children living with lone parents (mainly mothers) reached 15 per cent in Britain in 1989 – double the percentage of 1972 – and 37 per cent of women with a child under five were working full- or part-time. (These statistics are drawn from HM Government's statistical publication *Social Trends*, for the year 1992.) Young people, despite unemployment, have greater spending power and earlier autonomy than their parents had at their age. There is more open acknowledgement of homosexuality, with more lesbians and gay men forming households. A rapid proliferation of consumer products has been accompanied by a culture of self-expression and individualism that coexists uneasily with the family values proclaimed by successive governments.

The Gender Politics of Food

The resulting tensions were evident in the conversations about food and cooking we had with microwave users. The microwave oven was clearly implicated in change, appearing as both a source of stress and a solution to stress. In 1991 we carried out semi-structured in-depth interviews involving a total of 20 people in 13 households in South-West England using microwave ovens. The individuals interviewed crossed the social class spectrum in terms of employment category. The households were also observably different in their level of prosperity and comfort – rang-ing from relative affluence to relative poverty. Among these households was a house shared by students where we interviewed two (one male, one female). There was an elderly man living alone. And we also interviewed both partners in a lesbian couple forming one household. (The families photographed in the following pages were not among those who were interviewed.)

Other sources too inform this chapter. Our interviews with women and men at work, reported in previous chapters, usually included some discus-sion of their own use of the microwave oven. As an additional supplement to the qualitative household interviews we sent postal questionnaires to a

sample of microwave owners obtained from a retailer's warranty records.
We received 34 usable replies.

We particularly ensured that half of this small sample of households
comprised heterosexual couples living with young or mature children: we
draw on these ten in the section that follows. We interviewed all the
female partners, and in four cases the male partner was present for part
of the interview.

Among the heterosexual couples in the household interviews we found
evidence of considerable tension and stress over the provision of food.
The main contradictions people were handling were four: money versus
pleasure; nutrition versus time; togetherness versus individuality; service
versus fairness. An overriding concern was the creation and sustenance
of a good standard of living for the family. Underlying rocks threatening
to wreck the project of a quality life were not only lack of money or
pressure of work but also the inequalities and unfairnesses of heterosexual marriage or cohabitation.

First, *money and pleasure*. This was a time of high unemployment and
rising mortgage interest rates. People worried about maintaining the
household on their weekly budget. They felt guilty about extravagances.
Yet they saw expensive food as a source of pleasure that could heal
soreness in relationships, lift daily life above drudgery, and make
marriage or family existence the condition preferred by all household
members to other conceivable situations. Among our small sample of
couples, money was most frequently shared, with both partners free to
write cheques or use their plastic cards without reference to the other
except for major purchases. For all but the few well-off households,
however, careful management of the shared expenditure was important.
It seemed that women tended to see themselves as the more responsible
partner in this respect. One marriage was on the rocks due to the
husband's reckless expenditure. Another wife, Laura Castle, told how
her husband Matt just came back one day with a new television, 'and
I gave him a black look and said "Erghh! well how much is that gonna
set us back then?"' She added that for her part she would have been
quite happy with 'a clapped-out black and white, you know, to save on
the licence bill and everything'. Sheila Eaton joked about her husband
having bought their microwave at the most expensive department store
in town, the last place she herself would have chosen. 'No doubt,' she
said caustically, 'it was the first store he parked behind.'

Tensions between the partners also emerged in descriptions of their
shopping for food. For example, Laura Castle had had to persuade Matt
to scale down from the luxury foods he was tempted by when shopping.
'He's getting better now, say, buying pork chops instead of steak.'
Lesley Whelan and her husband Roger would do their Saturday supermarket trip together, but

erm – he tends to go for more expensive fruit than I will, so I'll go round and get the greens and potatoes and carrots [in a mock-sensible tone] in the trolley. Meanwhile, he'll be getting the grapes and maybe slightly more exotic fruit than I would. So, he stocks up the trolley with goodies and I get the basics, really.

Because her husband could not, as Lesley saw it, be trusted to think about expense, she retained the main responsibility for shopping as well as cooking.

Money, of course, cuts across the nutritional issue too. Lisa Howard felt she would like to buy more organically grown vegetables, but could not afford them. Many people had in mind 'treats' put beyond their reach not only by health and time considerations but also by limited finance. Some women, like Jan Ottley, longed to indulge themselves in pleasurable cooking. She enjoyed browsing through colourful pictures of rich dishes: 'I've got a thing about cookery books'. Joan Patterson would take time off from work before Christmas to have the pleasure of baking and cooking special foods for a truly festive occasion.

Secondly, people worried about *nutrition*. They felt that not only money but also, and particularly, shortage of *time* in their busy lives worked against their ideal of providing themselves and their households with a healthy diet. The people we interviewed felt they had changed, along with many of their acquaintances, in becoming more conscious nowadays of health and safety in shopping, cooking and eating. Publicity given to the risk of high-cholesterol foods, of harmful bacteria in chicken and eggs, the campaign against chemical additives and growing awareness of the damaging effect of pesticides and fertilizers – all this was adding to the work of whoever provided food to the family.

Jan Ottley for example said that today for her health was the main factor in food choice. It had not been like that when she first married. They had happily eaten chips, burgers and packet chow mein. Today, if she permitted herself commercial ready meals at all she would buy them only from her favourite supermarket, where she could be sure of 'nice healthy food – it's not like the normal crap when you normally buy these things'. Joan Patterson too had turned to brown bread, yoghurt and muesli. No more Sugar Puffs for breakfast in the Patterson family.

Though none could avoid the new health consciousness, it was not every household that had the time and energy to carry through a commitment to healthy eating. We heard several guilty expressions of defiance. For some families the presence of a microwave in the kitchen had been allowed to legitimize the purchase of more instant microwavable meals. Leonie Evans had recently returned to full-time paid work after a period at home with children. Jack Evans limited his input to a bit of 'help' on a Saturday. She was therefore cutting all the corners she could to deal with the double burden. She felt embarrassed and uncomfortable in the interview to admit that she was no longer cooking all the family food herself. Jack, however, had no inhibitions about saying how domestic

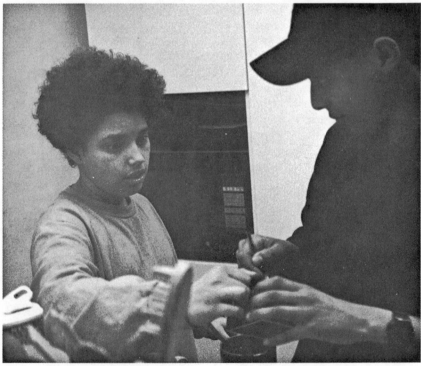

practice had adjusted to her desertion. They had bought a washing machine and tumble-drier, reflecting 'the change in our lifestyle with Leonie going out to work'. He said to her: 'I think your attitude, with shopping and buying pre-cooked meals was, "I haven't got enough time, so we've *gotta* have these things, and so. . .".' 'Yes,' confirmed Leonie, 'and so the children could get things for themselves.' As a result she limited herself now to making sure the teenagers received a daily token of protein, fruit and vegetables. The children however were overweight and Leonie's casual approach to nutrition (Jack, blamefully: 'You'll say in the morning to Christine, "Do you want a cornish pasty at school today?"') joined with many other disagreements to become a source of visible tension in her relationship with Jack.

Though we came across one family in which it was the man who had become a convinced vegetarian and environmentalist, and had taken over more and more of the shopping and cooking, in the main it was women who worried most about healthy eating. Instant meals, ideal micro-wavable food, were seen as unhealthy options. Lesley Whelan for instance said: 'Roger will go to Marks and Spencer's and come home with three or four easy meals, flan or whatever, pasta or something, a whole batch of easy-to-heat-up meals.' She on the other hand tended 'to go for healthier food, and avoid that temptation in selecting *for* him'. It was one more responsibility for women, experienced in some cases as quite onerous, to discriminate between the food advertised and offered in the shops and the food that would be best for their families. This often came down to a choice between fast, often microwavable, food and food that took longer to find, cook and serve in a situation where women's time was under heavy pressure.

Third, *togetherness versus individuality*. Growing children, working spouses and active lifestyles were fragmenting the experience of cooking and eating. Sometimes individual wishes were now being catered for in a way not possible before the development of a market in instant meals. Thus Sheila Eaton would sometimes now provide a treat for her husband, aside from the meal provided for the children and herself, in the form of a double-portion instant Indian curry. Jan Ottley would give her meat-eating sons packet beef lasagne while she herself would have a vegetarian one. Leonie and Jack Evans found they were eating separate meals during the week. The girls would come home from school starving hungry and get themselves something to eat before Leonie could get back from her job to cook for them. The same went for Jack. And Leonie herself might eat out on her way home. The 'family meal' was with difficulty, given the children's independent activities, defended as a weekend 'must'. The last bastion the Evanses would allow to fall was Sunday lunch.

Although there were expressions of nostalgia over the loss of family mealtimes, some people we spoke with had mixed feelings, for they also felt that the insistence of their own parents on patriarchal commensalism had been oppressive and was now outdated. They asserted their children's right to choice concerning food. The most extreme attempt we heard to permit individual choice in food, and freedom in mealtimes – that is, an enhanced quality of *individual* life, while sustaining in parallel a practice of quality home cooking – was that of Sheila Eaton. She said: 'I think the only day we ever sit down and have a meal together at the table is probably Christmas day'. To accommodate a fussy husband (who would not eat fish or stew), a living-in niece with baby, two picky teenage children and a little girl of five, she was regularly home-cooking a meal with seven individual variations (not to mention meat for the two dogs) and leaving it plated up under clingfilm to await their individual arrivals. One at a time they would pop their personalized plate into the microwave.

> *Sheila:* I mean to say, obviously, they've got their own likes and dislikes. . . It's like, now probably tomorrow night I shall make curry. Now there'll be hardly any curry in Debbie's, and obviously I shall take some out for the baby so there'll be no curry powder in that obviously. And then it'll be slightly stronger for Seb, the eldest one, because he eats very plain food. He doesn't like onions, anything strong flavoured, cheese, you know. And then it'll be stronger again for Doug. I'm not sure about Marline yet, how strong she likes it. The other one I could chuck the curry container in. And what I'll do is, you see, the microwave will come in use tomorrow. Because I shall prepare the meal, then I shall start adding the curry powder in degrees and dishing it up, per portion.
> *Interviewer:* That's quite a lot of organization you've got to manage then?
> *Sheila:* Yeah, terrible really. Yeah. [*Laughs*] Now you know why I'll only chuck in a jacket potato for myself.

Sheila's friends and neighbours, discussing her in the corner shop, were both admiring and appalled by her strategy.

> *Betty:* Sheila's fascinating isn't she with all her plates lined up [*laughing*].
> *Peggy:* I can't get over what she cooks [in her microwave]. She cooks everything in there.

Some felt the quality of Sheila's food failed despite these heroic efforts, because she was simply, when all was said and done, a bad cook. A woman producing badly cooked meals was a woman failing to serve the family – especially *him* – in the proper way.

> *Betty:* If I was to put that meal on the table I reckon he'd throw it at me. I mean if we're gonna have a fry-up, we like it put *fresh* out of the pan and on to the plate. I was horrified what she done, honestly. I've never seen anything like it.
> *Peggy:* What she done?
> *Betty:* Well, she'd fried in her dinner hour [i.e. ahead of time], and she'd got it on the plates all lined up. I mean she'd a *fried meal* and *she warmed it up in the microwave*. . . I wouldn't of ate it. But then we're more for home cooking. It's the way we've been brought up.

If women accepted the break-up, for most purposes, of the family meal and were resigned to using some commercial instant food, they nonetheless regretted the pleasurable occasion of eating in intimacy, the chance 'to sit around and talk, you know, with a bottle of wine. . . Life seems to be in such a rush.' Women longed to cook specially for their partner and eat alone with him, *à deux*. This ideal had become even more unrealizable when children came along. A young child usually meant falling back on unsophisticated foods. The 'family' meal, considered important in cementing the relationship between father and children, had the contradictory effect, regretted by the woman, of destroying moments of closeness between her and her partner: quality time with sexual meaning spent over pleasurable food. Leonie Evans, mother of two, regretted the days when she and her husband Jack used to 'spend more time eating proper meals rather than rushed things'. Jack recalled, 'We would spend more time preparing it, thinking about it'. 'Yes, yeah,' added Leonie, 'and we'd always, we'd always eat together then.'

Perhaps the most tension emerged on the issue of *service versus fairness*. We asked all informants about the sharing of activity in their households, the gender pattern of their location in the practices of shopping and cooking. In order to play its proper role in the traditional family, food should be cooked and served by the woman/wife/mother. Yet if she were now working outside the home, and to some extent even if she were not, today's ethic of 'fairness' and 'equality' proposed another norm. The woman *should* no longer be a servant to the family, men and children *should* help. And, to be practical, if they refused their share, either the quality of life must fall or the woman must be grossly overworked. Either way she felt a confusing mixture of guilt and anger.

Two women clearly still did everything in the home. Betty Vincent was one. Her husband suffered from anorexia nervosa. Betty ran his business and did a cleaning job on the side to generate a little income of her own, to spend not on herself but on her grandchild. She cooked for this man who refused to eat and for an adult daughter who grabbed a bite when passing through the family home to put her clothes in the washing machine. Finally Betty took care of her own housework late into the night. 'When I go home evenings I do things like ironing, things like that. Until I drop on the floor [*drily*]. And I can't sleep at night [*chuckles*]!'

Two of the ten men among these couples running family households (Adam Howard and Raoul Daws) could be said to have been fully sharing the housework and childcare. For most of these women, as we saw other studies to have found, their partners' contribution is modest – not sharing but 'help'. The help women get they welcome. Laura and Matt Castle are symptomatic. 'My husband is incredible. He does give an awful lot of help around the house.' Sometimes even a man's simple tolerance of lowered standards is interpreted as helpfulness. Thus Laura said she no longer ironed the family wash. But Matt was helpful in accepting this and was even prepared to iron his own shirts – the only

item where the external world, it seems, imposed the maintenance of a non-negotiable standard.

Women often felt bitter that their partners did not offer more help and that they did not clean, cook or care as well as was desirable. Marion Daws said: 'if he does put the hoover round I still feel he hasn't done a good job. So I. . . have a good old blitz. . . The main cleaning, down to the bottom of the gloves, is me.' 'Sometimes,' said Jan Ottley, 'I'd rather he didn't bother.' When men do 'help' it often releases the woman, not for leisure to spend as she pleases but to do other work. Thus Matt Castle would look after their daughter on a Sunday morning so that Laura could 'shut herself off in the kitchen' and cook the Sunday roast.

Even the men who gave considerable time to other aspects of housework drew the line, their wives said, at bathroom and toilet cleaning. All 'women's work' is demeaning, but, within the whole, tasks have varying value. Kitchen work is lower in status than living-room/caring work. Bathroom/toilet work is lowest of all. Women resent men's drawing the line. In a painful interview with the Evanses the following exchange took place:

> *Interviewer:* Who cleans the bathroom and toilet?
> *Leonie:* Well, that's without doubt always me. Nobody else will clean the bathroom and toilet.
> *Jack:* [*half joking, half needling*] Maybe you ought to ask who always washes the car!
> *Interviewer:* [*to him*] Do you wash the car?
> *Both:* [*laughing*] Yes.
> *Leonie:* [*invoking fairness*] He always *uses* the car.

There seemed to be a pattern whereby women and men had looked after themselves before marriage, had started married life intending to share the domestic labour, but had gradually (notably with the onset of parenthood) slipped into sex-stereotyped roles of serving and served. Pressure to earn and to improve the quality of life in the household had been a factor driving the couple into traditional roles. Jan Ottley said of her husband Maurice:

> he'd have been the one happy to have stayed at home with the children and just, you know – But it's like you're always [*quick, insistent tone*] push, push, push. And I felt like, you know, 'Do this, do that, go on, you know. Don't you think you want this wonderful four-bedroomed house with all the trappings?'

Women, Men and Microwave Use

In some cases technology helps men evade housework. Maurice Ottley, who had no time for labour-saving appliances in general, made an exception in the case of the dishwasher, a fact that Jan did not hesitate to ascribe to his self-interest: washing up by hand had been his contribution to the housework. What of the microwave oven?

We found a marked distinction being made, both conceptually and in practice, between time-consuming 'real home cooking' and rapid 'food heating'. Real cooking involved selecting good ingredients, following recipes, allowing food time to cook slowly and develop flavour, and employing skills – often learned from mother or grandmother. While the microwave found a ready role in, indeed prompted, a practice of food heating, real cooking did not preclude using the microwave oven. As we saw, Sheila Eaton was portioning up for the microwave what, by her own lights if not in the assessment of her neighbours, was quality home-cooked food. We found however that, even if the microwave was being used to serve up reheated real cooking, the actual preparation of this food was being done in almost all cases by the woman and in a conventional oven.

We saw in Chapter 3 how the home economists at Electro created recipes to attract serious cooks to microwaving. They are supported in this exercise by home economists in the teaching of domestic science in higher education, and also by a small coterie of cookery writers developing a *haute cuisine* of microwaving for the discerning and quality-minded housewife. A home economics teacher we interviewed felt that the microwave oven manufacturers and those branches of the food industry developing the ready-meal market were combining to 'usurp the traditional housewife's skills'. 'The number of people who choose basic ingredients these days is very few in comparison with the number of people who are choosing ready-prepared foods.' She felt that doing real cooking today had become a hobby for a minority of women. 'In the way people take up quilting, somebody might take up cooking in a big way. It's at that sort of level.' She saw the practice of proper mealtimes as being eroded. She and her colleagues tried to teach microwave cooking as an additional skill, rather than as a cop-out from cooking. But this was clearly an uphill struggle.

A well-known cookery writer agreed. There was an irreversible trend away from the true family meal. 'I don't like to think that,' she said. 'I don't want to think that. But it's true. . .' Both adult and young family members today had so many activities. 'How can you possibly arrive in the same place at the same time to have a meal? I think that's one of the things, I think it's our society, our sick society, that's forcing this. . . I think if they all ate *together*!' She feared the forward march of microwaving and instant eating was unstoppable, and the art of cooking might well die within ten years. Her life project had become the raising of microwave use to real cookery, the enrolment of women into the practice of microwave cooking without loss (and indeed with some gain) in standards.

According to available statistics, these microwave propagandists are failing in their attempts. A survey of 1040 housewives in 1991 found only around one-quarter of microwave owners using their oven to 'cook' cakes, puddings or meat. (No reference can be given here for reasons of anonymity.) By contrast 86 per cent used them for defrosting, 82 per cent for reheating home-made food, more than half for reheating ready-meals and tinned food. The nearest most people came to cooking from fresh ingredients direct in the microwave was cooking fresh vegetables, for which microwave has won itself a good reputation (43 per cent did this).

A research manager at the Consumers' Association, in interview, confirmed that the Association's many surveys showed that

> people use microwave ovens for very simple tasks. The top three in popularity are cooking vegetables, reheating drinks, soups and so on, and reheating convenience foods. The fourth category is reheating food they've cooked themselves. Cooking food from scratch is way down the list.

The implication is that although people value a 'browning' facility (24 per cent like this idea), full combination cooking is a rarity even among combination oven owners. The Association's research also shows that microwave owners seldom use programming and automaticity. Auto-defrost, auto-reheat and auto-cook facilities remain under-used in most cases.

As it happened, only three of the households in which we interviewed had combination ovens: one belonged to a woman sharing with her lesbian partner, and two were in pub kitchens. From their own description of their use of the microwave it appeared in any case that anything other than basic facilities were seldom used. Our ten heterosexual family households fell into three categories. First there was the minority (two) who used their microwaves in an experimental way, embracing their potential, incorporating them into their cooking practice and cooking recipe dishes from scratch. It was Sheila Eaton and Joan Patterson who fell into this category. Only one-third of the respondents to our postal questionnaire said they 'cooked meals or more elaborate dishes from a variety of ingredients'. And for every person who replied that they would like to have additional features and facilities to those supplied on their current models there were two who 'weren't bothered'. Women were even less interested in upgrading than men.

The two other groups used the microwave only for simple heating or the cooking of one item at a time, such as carrots or jacket potato. What differentiated them was that one group used it regularly, the other rarely. Perhaps the typical microwave user is the one who uses it frequently for very simple tasks. Six out of our ten family households and two-thirds of the respondents by post fell into this category. The Castles and the Ottleys are good examples. Laura Castle used her machine often, even, unusually, for joints of meat. She had simply allowed culinary standards in her home to fall to cope with hardship, a baby and now a pregnancy. They just 'got used to having things soggy'. In the Ottley household the microwave was used every day for such tasks as heating the children's chocolate drinks and defrosting bread. Jan used it for frozen ready-meals in a cavalier style, defrosting and cooking in one move. If you were going to switch the thing on 'you might as well zap it through'. Though now and then she aspired to doing something more fancy, on the whole she felt satisfied she had this machine in an appropriate perspective. 'I get very lackadaisical. . . I think I'm not gonna be a slave to it. I'm not gonna be taken over by it. It's only a machine, so if I want to abuse it I will.' In such households the microwave had become a useful, routinely used, heating device, something that had simply 'found its place between the freezer and the cooker'.

In two households, however, not only were the microwave oven's full range of features not used but the basic oven itself was forgotten for days at a time. This also applied to 13 of the 34 postal respondents. One of the under-using families was the Howards. Lisa had started out keen. She had even had a go at cakes, but they had turned out to be rock cakes in substance as well as name. Like many users, after this initial failure (sometimes it is gungy rice, sometimes exploding eggs), Lisa let the microwave go hang and returned to her beloved wood-burning stove, the warming heart of the house.

Men in these families, whether they had begun as the enthusiasts, bringing home a microwave on their own initiative, or as the sceptics, expressing suspicion and doubt, had in the main learned over time to be adequate zappers. (This is the term used by the Consumers' Association to intimate the kind of person found in their surveys, usually a man, who microwaves without reference to instructions, using the maximum power setting and guessing the time.) They were certainly more likely to use the innovation than they had been to use the conventional oven. We were told of one woman who reported that her husband had less objection to using the microwave 'because he does not have to bend down' in the way one does to open the door of a conventional oven. 'I think it's bending down with a cloth and getting things out of the oven is an association with his mother,' was our informant's view. Standing upright at a worktop does not carry that demeaning feminine connotation. One man in our sample noted his own readiness to stand at an eye-level grill in contrast to a table-height hob. These things confirmed the perception of a home economist we spoke with, who felt the microwave was more popular with men than the conventional oven because 'it doesn't make them a sissy, if you like. It's modern, it's a technological thing, therefore it isn't sissy to be the mother. They're not replacing mother at all, they're just assisting.' There is still a touch of gender ambiguity about the microwave oven. Despite its fall to the feminine environment of the kitchen, it retains just a whiff of aftershave.

These men mostly knew how to get out of the microwave what they personally needed: a cup of coffee, the plate of supper left by an absent wife. None, however, were microwave cooks in the real sense of the word. Adam Howard, one of two men who really cooked, not only for himself but for the family, disliked the microwave. 'I hate it, I hate the things. . . You don't know what it's doing.' Even this difference of opinion was grist to the mill of domestic tension. 'Lisa really wanted it. I didn't want it.' Lisa tried to defend herself, but Adam continued to reiterate disgust. Some of the men continued to sham incompetence. Of Simon Eaton his wife Sheila said he would sometimes put something in the microwave if she was busy but 'he'd always ask how long'. Her niece Marline protested, 'He didn't ask *me* today how long his pasty and beans would take.' Sheila laughed and said, 'That's probably because it was you.' She was emphasizing that she understood Simon's behaviour to be an expression of their own relationship as a couple.

The microwave then is associated with a greater tendency for men to reheat food but not to cook food. It is linked with a greater likelihood that men will help themselves to food but not that they will serve others. The microwave is an inseparable part of a set of changing circumstances – in employment, in the market, in the family, in people's consciousness. But it steps into an existing gender context and helps to shift the masculine/feminine relation into a slightly adapted, modernized, mode without really transforming the important imbalance, the differential value, the hierarchy that characterize it.

In the course of half of our ten family household interviews – and it was particularly obvious when the man was actually present – we sensed considerable friction between the partners over the apparently innocent food matters discussed. Between the other five couples too mild differences, disagreements and disappointments were expressed. It was striking to us that such tensions did not appear to exist in the five households that did not have a traditional family structure involving heterosexual couple plus offspring. For the students in their collective house, food had hardly any political significance: it was a straightforward matter of minimal provisioning. Nobody depended on service or help from anybody else, there was no cause for guilt or blame.

Of the two young lesbians who had set up house together, both of whom we interviewed, Kate had taken on all the cooking, while Anita reciprocated by washing up and cleaning the kitchen. There was no obligation or expectation involved and no differential value ascribed to the chosen roles. It was simply that Kate had always loved to cook. She had surprised her parents by asking for a microwave oven for her 21st birthday, and had since upgraded to a combi. She loved to prepare delicious meals for her own pleasure but clearly also enjoyed 'treating' Anita. Kate, and the 71-year-old widower, Mr Brown, who lived alone, were the two most whole-hearted gourmets we came across. How it had been when Mrs Brown was alive it was impossible to know. (Mr Brown said they had 'shared' the cooking.) Certainly now, with only himself to please, he was developing an ever-extending microwave repertoire of which he was proud and in which he delighted.

Our sample is small and not necessarily representative. It lacks for instance childless heterosexual couples. Yet it is clear that the microwave oven enters into existing food and cooking practices that, as other researchers have suggested, have a very particular and pronounced symbolic significance within the heterosexual relationship in the context of the contemporary nuclear family household (Delphy, 1979; Murcott, 1983b).

The Gendering of Technical Competence

One theme in the interaction between women and men in our ten family households was the microwave itself as a *machine* and the understanding of engineered artifacts, as opposed to cooking, one had to have to use it confidently. Microwave ovens have something in common with all modern technologies and something that sets them apart. Like other artifacts they are endowed, from one model to the next, with more and more features and facilities, more and more complicated controls. Unlike other artifacts they generate microwave radiation, on the safety of which there are differing views.

There is undoubtedly still, after a couple of decades of microwave cooking, a deep-rooted fear of potential radiation hazard. Among the more than one thousand 'housewives' surveyed for Electro UK by their advertising agency a quarter felt 'they are too dangerous because of the radio-activity' and a further quarter were undecided about safety (no reference can be given for reasons of anonymity). Among the panel discussants organized for the Association of Manufacturers of Domestic Electric Appliances safety was a significant barrier to purchase for non-owners of microwaves. 'There was much talk about radiation, leakages. . . There was a feeling that microwave oven technology was just too young for the full effects of usage to be known or researched, and this made many nervous' (Higgins and Almond, 1989).

Both sexes admitted to us some concern on the radiation score. 'What worries me is how do you know if there's a leak.' 'I never stand too close.' 'I'm not totally at ease with the idea. . . It's an unknown factor, all those molecules jumping up and down, getting themselves warm. It doesn't seem quite natural, really.' Many people had such lingering doubts: could microwave radiation cause cataracts, sterility, produce miscarriage? But they just got on and used the thing anyway. 'I don't think if I had a baby I would [use it], but it's only us two and it doesn't seem to matter. We're almost downhill now.'

On the second issue, that of the challenge of understanding and controlling the technology, there did appear to be more women than men who lacked confidence. Tess Barrington admitted, thinking of the time she first got her microwave,

I was just wary of it because I thought, well, it was alarming really, doing this or that. And I thought 'I'll never get the hang of that!' I tend to be very conventional, I tend to shy away from things that are automatic and modern, technological. . . I think I'm a bit taken aback at what they can do now. But I'm not that technical. It would just alarm me.

Often they compared themselves unfavourably to their partners in this respect. Joan Patterson said she wouldn't have been the one to attempt to wire a plug for the microwave. 'Certainly not me, I haven't a clue. Simon or one of the children would do it. . . I don't think I'd be inclined to touch it at all.' Lisa Howard's husband too would be, she said, the one to install new appliances. 'He does, he did [*embarrassed laugh*]. I'm not very good with electrical things.'

At one level, women in speaking about themselves are reproducing a widespread stereotype portraying women as technologically incompetent and prone to technofear. (In reporting their observations we are in danger of doing so too – such is the mutual shaping of the material and the representational.) In fact, though the men we met did not admit to *fear* of the microwave, some clearly understood its functioning less well than women supposed them to. Laura Castle saw through her husband's bluff. 'Matt will try and *pretend* he knows a bit more about it,' she said. And clearly many women are practically very competent at controlling the microwave-in-use, whatever the distance they feel from technology-as-engineering. On the other hand we know from other research that men are more often than women self-acknowledged technophiles (Cockburn, 1985; Hacker, 1989). The AMDEA study on attitudes to microwaves, reported above, found 'the men seemed interested in any new technological developments and for them it was almost a criterion in itself to have "the latest" model, regardless of any particular suitability' (Higgins and Almond, 1989).

Again, as with feelings of anxiety around cooking, anxiety about technology was strongly associated with *interaction* between women and men. A woman can feel put down by a man. And indeed they sometimes are: Betty Vincent's husband told her to take back to the shop the microwave she had chosen, and exchange it for a simpler model without programs and with dial controls. 'He knows I'm stupid. . . He's always telling me I'm daft.' A woman feels vulnerable to criticism. Lisa Howard for instance let Adam choose the model, so that 'he can't moan at me if it's wrong'. And, she said, to avoid being shown up, she would wait for him to leave the house before 'having a play with it' to familiarize herself with the controls. She was not the only woman who, simply to avoid 'bother', had abdicated responsibility for learning. 'I tend to let him read all the instructions and then tell me.' One or two women indicated that their technological competence had either been impeded or eroded over the years of their marriage. Thus Marion Daws said of wiring a plug, 'I could do it when I was single, but I've forgotten how to do it now. I haven't done it for so long.' Kate, Anita's lesbian partner, was impressive in the sheer confidence and versatility she expressed with her microwave. She put it down to having 'grown up with lots of gadgets' – but it may also be relevant that she had had no male partner with greater technical confidence to drain away her own. The imperative of maintaining the proper authority of the male in the heterosexual marriage was absent in her case.

What these household studies affirmed more than anything was that technology as knowledge and as process is *a relation*, and in that relation among others subjective and projected gender identities are shaped. What kind of people women (and men) think they are, and who they are seen as being, has to do not only with the things they do and do not do in paid employment but also with the things they are seen as 'good at' in the home.

6

Gender: Making and Remaking

In Western culture, 'technology' is surrounded in mystique. When we hear the word we think first of the more complex, powerful and even dangerous inventions: the computer, the space satellite, the nuclear power plant. Technology, however, properly has a more everyday meaning: the knowledge and practice of doing, making and producing. Technology involves continually evolving new ways of doing things, and the aids to do the doing: tools, appliances, machines.

By looking perceptively at technological change, therefore, we can see social change – for doing, making and producing are a fundamental part of social existence. In studying the innovation of microwave cooking we have seen technological production at two levels. We have seen ordinary people in everyday life producing their meals with cookery knowledge and technical aids such as wooden spoons, pressure cookers and microwave ovens. Secondly, we have seen people producing the appliances with which people will eventually cook, using engineering knowledge, the science of home economics, manufacturing technologies such as intermittent flow assembly lines, and retailing technologies such as Electronic Point-of-Sale computer programs. For an innovation at the first level to come about, new activity must take place at the second, and also of course in many intervening and related sites, such as the food, container and transport industries, that we saw in Chapter 1 go to make up the actor-world of a new technology.

The social relations within and connecting these sites we have called *technology relations*, because they are the relations in which the social construction of a technology occurs. To call them technology relations, however, is not to describe them exhaustively. They are also relations of class. To take just two illustrations of this: the owners of capital invested in microwave production and distribution seek the cheapest and most effective labour for the various tasks of design, manufacture and selling; and their calculations concerning the new technology involve the class relations of the consumer market. Technology relations are also relations of race: we saw, for example, something of the interactions of the Japanese and British actors in the microwave innovation.

Gender Relations Shape Technology

In a complex, cross-structured manner, however, technology relations are

also and inevitably, *gender relations*. Inevitably, because gender is one
of the major structures of the social order and gender relations are
found wherever people are found. This applies, contrary to popular
usage, even when the people are all of one sex. Men, for instance, in
an all-male environment relate to each other in ways that express forms
of masculinity, every bit as much as they do when relating to
women.

In the foregoing chapters we have seen something of gender relations
as they shaped the microwave, gave this technology a particular social
identity. Microwave cooking was an idea that had some of the 'multidirectionality' we have seen characterizes innovations. That it took
certain directions rather than others was due to gender relations. Of
course such considerations as the price of materials and transport, the
buoyancy of regional consumer markets, were influential too, deciding
many of the microwave's features and trajectories. Even within such
factors, however, usually designated 'economic', gender, race and class
relations are at work. A corporation with world-wide production,
for instance, locates its plant in the light of labour cost calculations
that take account of the availability of Third World women workers
– who, due to the unequal relation of men to women and of industrialized to non-industrialized countries, may be hired at a lower wage
and on terms more advantageous to the employer. Besides, gender relations are not absent even from such factors as the cost of materials
and transport, through the labour-cost component involved in their
price.

In Chapter 3 we saw the way in which a microwave oven was
designed with two futures in mind: the immediate prospect of the
process of manufacture, and the more distant prospect of purchase and
use. The gender relations foreseen in each of these processes entered
into decision-making. First, the availability of male and female labour,
seen as embodying particular gendered strengths and weaknesses,
shaped the decisions of Electro's Production Engineering Department.
Second, the company recruited (feminine) home economics knowledge
to supplement the (masculine) engineering knowledge of its Design
Engineering staff with the effect of gendering the artifact in
appropriate ways for its imagined end user. The women in the Test
Kitchen were able to a certain degree to use the interpretive flexibility
of the microwave, lending their influence to support the development
of the combination oven, reflecting the preferences, as they saw them,
of the traditional woman cook. The range of ovens, however, includes
increasingly technologically sophisticated machines with a high degree
of automaticity, reflecting the enthusiasms of (male) engineers, for
which a (gendered) user had yet to be fully visualized and perhaps even
created.

In Chapter 4 we saw the social, gendered identity of the microwave
oven modified as it entered the retail trade. The oven itself, or rather

Electro's range of microwave ovens, had temporarily achieved closure, entered production and arrived on the market. The artifact itself could not now, in the short run, change its material features. Yet sales practice could even now embellish or alter its meaning. An identity is projected for the artifact by its positioning in the store and also by advertising, point-of-sale material, instruction booklets, the way it is spoken about, the sales pitch. Here again gender plays a part. We saw the microwave shift from being a more or less masculine engineered product to being a 'family' white good, on the way to its anticipated home, a feminine kitchen environment. Once there however it retained a trace of its former masculinity, less alien to men and boys in the household than the gas cooker or electric stove it supplemented.

A modicum of flexibility is retained right up to the moment of use and beyond. What we are interested in is not the artifact in and of itself but the new knowledge and practice of cooking it suggests and enables. We have seen that in the highly gendered arena of the domestic kitchen the microwave is shaped by the relations into which it is obliged to fit. Often it is put to a use that is not quite the one its manufacturers intended for it. Even sophisticated models are often used (particularly by men) in undiscriminating 'zapping' mode, and, in combination ovens, microwave energy and conventional heat are seldom in practice activated simultaneously in the way intended by the designers. The user's choice to neglect expensive facilities s/he has purchased is, on the face of it, no loss to the manufacturer: the sale is by then complete and the profit banked. It is an own-goal scored by the user. In a clumsy way, nevertheless, the learning process of the manufacturer, via the customer complaints departments and service engineers, leads to reopening the closed design and reshaping the artifact.

Gender is unavoidably at work in the whole life trajectory of a technology. In this final chapter we must turn attention to the inverse process, and draw out from the story the ways in which technology can be seen to enter into and help construct divergent gender identities.

Identity: the Given and the Chosen

At the outset of this book we defined 'identity' as having two aspects: subjective identities as lived and experienced by the individual, and projected identities generated in culture and offered, so to speak, to a myriad individuals who may or may not 'identify with' them.

Subjective identity is in some ways analogous to society: it is a kind of organization with a structure or patterning built in by past experience. Without such a demonstrable identity, a society and a person is unthinkable, literally a non-entity. Antonio Gramsci wrote, 'Each individual is the synthesis not only of existing relations but of the history of these relations. He is a *précis* of the past' (quoted in Rutherford,

1990a: 19). A woman's sense of herself as a woman, for instance, will be different after ten years in a convent boarding school than it will be after a similar period in a coeducational inner city comprehensive. The form of identity she emerges with may well limit the direction, scope and speed of possible change. The form differs both in a collective sense according to such categories as religion, class and ethnicity, and according to personal psychology (as individuals we may experience such schools differently). One way or another, at any one moment an individual's sense of self – who she thinks she is – has a historically constituted form that inclines her to certain thoughts and behaviours. It guides her choice of a job, for instance, her demeanour, her ways of relating to her own sex and to men.

As with social structure, however, the structure of an individual identity is continually subject to 'remantling' and change as time passes and experience accrues. The experiences that stimulate reorientation and redefinition of self may be those of a collectivity (black women in Britain in the 1970s, for instance), or they may be individual experiences (making a new friend, perhaps). The evolution of subjectivity may be joyful or painful. The pull that stasis exerts against metamorphosis is powerfully expressed by Frances Angela:

> I have come to believe that my subordinate identity, growing up as a girl on the margins of the working class, defies any ability to move completely beyond it. Casting off the logic of our pasts and histories that runs through our lives is not easily done. . . The subjugation of my body, emotions and psyche, the lack of opportunities in employment and education. These are the traces of the past that work even now on my mind and body, that have left their marks and scars on my mental and physical health. (Angela, 1990: 72)

Powerfully though circumstances may shape it, however, identity is never wholly 'socially constructed'. Experiences are always creatively assimilated by a unique person. It is this element of creativity in the growth of self that introduces, beyond the 'givens', something of the elective. Jeffrey Weeks elaborates this in the following passage:

> Identity is about belonging, about what you have in common with some people and what differentiates you from others. At its most basic it gives you a sense of personal location, the stable core to your individuality. But it is also about your social relationships, your complex involvement with others, and in the modern world these have become ever more complex and confusing. Each of us lives with a variety of potentially contradictory identities, which battle within us for allegiance: as men or women, black or white, straight or gay, able-bodied or disabled, 'British' or 'European'. . . The list is potentially infinite, and so therefore are our possible belongings. Which of them we focus on, bring to the fore, 'identify' with, depends on a host of factors. At the centre, however, are the values we share or wish to share with others. (Weeks, 1990: 88)

Gender is only one, albeit an important one, of the many aspects of identity that structure our personhood. Jeffrey Weeks names some others above. Yet we never experience any one of these except through the others. The 'I' is never only a man or a woman; always some particular kind of woman or man. Nor is gender ever a simple positioning on one side or the other of a masculine/feminine divide. Apart from outward articulations of gender with identities such as class and ethnicity, there are inner articulations: submissive femininity, autonomous womanhood, lesbian identities.

The individual's material positioning in what we have called the gender pattern of *locations* – the work she does, the organization in which she does it, the skills she has qualified in, the responsibility she takes on and the particular share she contributes to the activities of the household – is an important factor disposing her gendered self-identity. The complementary shaper of a person's sense of self is of course the battery of *representations* that reach her perceptions. A teenage girl, for instance, is continually receiving and responding to projected gender identities – what her mother or teacher transmits as proper feminine behaviour, the young women characters portrayed in TV dramas, in fashion ads, feminist magazines and sociological texts (if she reads them). Ideologies and moralisms of many kinds patrol the boundaries of identity. A woman's gender identity then is pulled here and there by material circumstances that locate her (let us say, her lack of wealth, lack of skill, the chains that bind her to the assembly line), by the way she experiences these locations (boredom, aspiration) and by voices that speak to her (she heard someone say 'there could be more to being a woman'). As we pointed out in the Introduction to this book, there is no hard and fast distinction to be made between the material and the representational, since each is responsive to the other. Always the individual negotiates creatively between the two, oriented by the personal psyche.

Technology: Inscribing Value in Masculinity

The women and men we met in the microwave world were evolving their gendered identities, in this way, disposed by location and proposed by the stream of images they encountered, but never altogether bereft of the possibility of choice. The microwave-world, which provided the locations we studied and the representations we heard and saw, was of course only a fragment of the life of each individual we met. Work on the microwave assembly line or in the electrical goods superstore only accounts for eight hours of the day. Cooking with a microwave oven occupies only a tiny part of the time and consciousness of most users. These women and men brought to the arena in which we interviewed them identities shaped and influenced by the wider social structures. The people we met in the workplace had come there bringing with them probabilities generated by their positioning in other contexts. The family, for instance. Nobody was unshaped by the prevailing sex-gender system which had laid out, like a hand of cards spread face up on the table, the possible and probable places of women and men in the family and the nature of 'the family' in contemporary society. The relative importance of men and masculinity we saw in this study was not an artifact of the microwave-world alone. The social processes of the microwave-world may have embellished and enhanced it, but the wider societal pattern disposed it.

Nonetheless, microwave technology relations may serve as an analogue for technology relations as a whole, and what we see of their part in gender construction may say something of more general significance. Doing, making and producing – technology in its general sense – are activities that clearly help shape our sense of self. Technology, even if we limit its meaning to tools, machines, artifacts, is an intimate partner in our daily lives. Our hands continually grasp it (ballpoint pen, steering wheel, hair-drier), it encompasses us (track shoes, central heating, shopping mall), it even takes up its place within our bodies (contact lenses, tampon, heart pace-maker). If we give technology its fuller meaning of knowledge and process, we find it, first, filling our thoughts and constituting our capabilities, and, secondly, filling our time and governing our movements.

The things we do and do not do, the technologies we are competent with and use frequently, the technologies we are intimidated by and leave to other people, partly shape our self-identity. We have shown in this study how women and men tend to differ in this respect. Technology is gendered. We collectively gender it, of course; but in turn it individually genders us.

What we saw in the life circuit of the microwave oven was a commonplace symbolic construction of the masculine and the feminine through technology relations (involving both locations and representations) as *different*, *complementary* and *asymmetrical*. Engineering competence was a fault line in the sexual division of labour. Men had jobs that involved it; women, with few exceptions, were located in 'not-engineering' jobs. Engineering was constructed as different from, and more important than, home economics – yet their complementary parts in the making of microwaving was acknowledged. Women, however, not men, filled locations requiring home economics knowledge. Engineering jobs paid better than those of home economists, and they promised better career prospects. Very important in producing and sustaining this material differentiation were representations, meanings made and purveyed by actors in the microwave-world, that mapped masculinity/femininity on to technology/non-technology and ascribed the pairs unequal value. Let's recall a few examples.

Sometimes it was technology itself that was ascribed the relative importance, as when Michelle, home economist at Electro UK, reported her colleagues as expressing 'awe and respect' for the 'machines and things that most of us don't understand' in the data processing department, identified with male 'whizz-kids'. Sometimes the technological artifact or technological knowledge suffered a relegation of importance because it

was associated with the domestic or feminine. In this sense the microwave oven had, in the Electro home economists' version of the story,
been under-promoted by the company's marketing team. 'The microwave
oven side,' Helen believed, 'gets neglected as far as promotion goes. It's
just one of the domestic appliances and not rated very highly. Hi-fis,
TVs, camcorders and things like that get priority.' We saw that in the
view of Dorothy, the senior home economist, the commitment to producing microwave cookery books at Electro had 'left a lot to be desired'.
She felt the microwave oven as an engineered product had been given
priority over the imaging-work on the microwave cookery process
without which, it could be argued, the oven itself would be valueless.
'The cookery books were developed in the past very, very fast,' she said.
The unwillingness of Electro management to invest in cookery book
publishing was, she felt, because the cooking process was seen as
feminine, and therefore, despite the objective purpose of the oven, of
less importance than engineering.

> It takes approximately a year and a half to develop a decent cookery book,
> if it's tested. Any cookery book, I mean the amount of test work, it isn't a
> case of a cake in the oven, 'Oh, it's taken four and a half minutes!' and that's
> it. It's the development of cooking as a *science*. It's not what most people
> think, *a housewife's job*.

In the same way, in the retail trade the fact that white goods were seen
as family or feminine artifacts produced a perception that therefore they
must be less 'technological' than brown goods. Thus the training
manager at Home-Tec said:

> what's strange is that when you actually look at some of the white goods
> available, you've got these super-chill refrigerators, ecologic washing machines
> and all the Aquarius and Zanussi stuff. It's microchip technology on a
> washing machine. So if you break it all down, a washing machine is probably
> just as complicated now as a hi-fi. But it's still this stereotyped image, if you
> like, [because] white goods is female and brown goods is male.

The distinction between technology and non-technology, the technical
and the social, inscribed with the meanings hard and soft, important and
relatively unimportant, interesting and relatively uninteresting, masculine
and feminine, was expressed over and again in our interviews. It was
especially pronounced at Electro UK, where technology relations ruled
and the consumer was still a distant prospect. Many casual remarks made
in passing in our interviews continually reminded us of the relative
appreciation of men and the masculine, relative depreciation of women
and the feminine. Tom, for instance, the most sympathetic to women of
the technical managers we met at Electro, described the way men were
invested with technical authority by both sexes. The assembly line
women, he said, 'don't like working for other women. Women often
prefer to work for men. Men probably prefer to work for men.' At
another moment Craig, a production engineer, gave an insight into the

relative value ascribed to courses at the college where he had trained. There had been mainly boys on his technical course, he said. But, 'if you go to another course, what we [called] *Mickey Mouse or social science courses,* er, they were all girls and weren't the boys.' This unselfconscious confidence of men in men, in the masculine and the technical, their undervaluation of the feminine, the human and the domestic, particularly but not only at Electro UK, were striking to us and often commented on by the women we interviewed.

To remain with Electro for a moment longer, the material (locational) gender structure current in the company's culture was *disposing* towards, and the gender symbolism (representation) was *proposing*, particular gender identities in relation to technology. To be masculine was to be technologically competent – that is competent either to deploy engineering skills or to manage engineers, to be relatively pro-active, project oriented, controlling. To be feminine was to have little or nothing to do with engineering, but rather to be dexterous and diligent, to know about people, domesticity and cooking, to be servicing, supportive and relatively available. Note that there is nothing intrinsically more valuable in one or the other of these lists of attributes. They are, however, ascribed asymmetrical value generally in our culture and specifically within the relations of Electro.

Of course, individuals of both sexes were already disposed to seek appropriately gendered jobs by their participation in the gender structures of the wider society and their receptiveness to the gender symbolism of the wider culture, projecting as it does more stereotyped than divergent gender identities. The situation in Electro simply confirmed these shaping influences, made the structure more concrete and the representations more imperious.

The individual is not clay, however. Each history is unique and, despite the powerful hegemony of the conventional gender order, people do make choices. Some individuals do respond to a contrary gender symbolism that exists in the interstices of the hegemonic culture, a subversive appeal to shatter gender stereotypes. It may be interesting at this point to return to two people we met at Electro who had independently positioned themselves in locations where their presence, the way they perceived their work and the subjective identities they were forming, were in some ways in contradiction to the gender relations of the company.

Innovatory Subjectivities: Wrestling with Technology

Karen, you may recall, was an exception to the prevailing sexual division of labour in Electro's production plant. She was a work study technician, employed in a junior capacity alongside production engineers. She had left school at 15, and drifted through a succession of predictable feminine jobs. She first tried her hand at hairdressing, then moved to sales assistant in a clothes shop, joined a catering course, did a stint in a fast-food kitchen and finally found herself seated at a supermarket till. When, fed up with that, she followed other local young women to the Electro assembly line at first she enjoyed the novelty. It was better, she said to herself, than being isolated at the checkout with impatient customers. But soon she got bored.

Karen had not been there long before she was promoted to 'junior leader' on the line. Then the post of work study technician was advertised. It was Tom, the equality-minded assistant production engineering manager, who encouraged her to apply and sent her on a short course in work study methods. By the time we met her, Karen was 22 and about to start a more advanced work study course by day release. The phases of her work began with the introduction of each new model of microwave oven. She would help design the tasks of the production line, determine their length and speed, 'balance' the line. The tools of her work study trade included a camera, a drawing board, calculator and stopwatch. She had by now become fully committed to work study and her horizons extended beyond Electro where she felt she was 'only scratching the surface' of the potential job. She thus had an ambition that does not occur to many women: to make a responsible career in a role at the heart of industrial production.

Karen was swimming against the tide, but she was not a total anomaly. There existed at least a concept by means of which colleagues and bosses could think of her. Because if sex-stereotyping was the norm, a new current, the notion of sex equality, was stirring in the world of work beyond Electro's gates. Little eddies were even felt within Electro, as some women spoke up for their own interests and Tom, at least, defended them. The language of equality allowed for the beginnings of a modified meaning for both woman and technology: a *technically competent woman*. It was in this cultural no man's land (and no woman's land) that Karen was cautiously constructing a gender identity. Cautiously – because, as we saw in Chapter 2, she hesitated to claim a technological nature for her work, which spanned the workers, the labour process and the product. Nonetheless, there was no doubt that she had stepped out of the feminine sphere within Electro and had intended to do so. When she was having that tussle over the expectation in the Production Engineering Department that she would make the tea, she had (she said) kept thinking to herself, 'I'm *technical* here. I'm *technical*. I'm not supposed to be making cups of tea.' She said: 'as a girl, I think, you've got a lot to prove. Women more than men have got to prove that they're not just here to answer the telephone.'

Her innovatory gender identity had created a degree of discomfort for Karen. She had spoiled her friendship with women on the line by moving up to this more highly valued, masculine role which involved her in scrutinizing their output and designing their labour processes. She it was who was often called upon to *locate* the sexes in their sex-typed positions, a function normally that of men. She would have to say, 'a girl can't do that job and a lad can't do that job. . . so you find you've got to put the girls doing the wiring and the lads doing the heavy stuff'. She had been instrumental in speeding up the line and generating more pressure of work than she had herself experienced while an operative like them. That this gender-contrary positioning she had taken up in relation to technology was putting her gender identity under stress was evidenced by the fact that she felt her marriage was holding her back.

The second person who comes to mind as having been in some manner generating a divergent gender identity in the context of technology relations is *Keith*, microwave product manager. While Karen was trying on a new sense of self as woman-in-a-man's-job, Keith was contesting the strongly projected technological masculine identity current in Electro. You may recall that he was in overall charge of the Test Kitchen and expressed himself more in sympathy with the female home economists and the interests of the domestic consumer than with his male colleagues and their entrancement with high-value hi-tech.

Keith had told us that his product, the microwave oven, had to fight its corner against 'a male prejudice against the product group'. In comparison with the 'glamour' surrounding the company's developments in artificial intelligence and liquid crystal display, boring old white goods tended to be 'the poor relative to some of these boisterous macho groups that are about'.

> The microwave group is a profitable group, and therefore we have respect because of *that* side of it. But in terms of the excitement factor or whatever, we're all keen to handle the camcorders when they come in, or the latest piece of audio equipment, or go down and play with the karaoke system. . . To try to persuade the men to come in here and look at our microwaves, it just doesn't happen you know. 'Oh, let's go out of our way to look at the latest microwave.' Just *not* interested!

He had no hankering to change his lot, and it was all one to him if his status in the eyes of male colleagues was the lower because of his location in the domestic appliance field. He was happy, in the way relatively few men are, to gain his sense of a professional self through the eyes of women.

> I always like to gain the respect and the opinions of my women colleagues I work with, and those sales people, the female people involved in my goods. I value their opinions more strongly than I do perhaps male colleagues' opinions. And also I find myself trying to look from a woman's perspective in terms of fashion or styling [of the product]. . . So I try to encourage that in some respects. I want our products to be better in the marketplace so that the sales are obviously greater.

Siding with women and women's interests, being 'a man involved in a woman's world' had become part of his conscious subjective identity, something he readily expressed about himself to others. In turn this identification was orienting his relation to technology. His view was that 'the whole consumer electronic marketplace has gone head over heels for technology for technology's sake', and he himself was supporting a turn to 'providing technology for the consumer's benefit' and to '*soften* the approach' too in the marketing of brown goods.

Keith achieved this only by turning a deaf ear to popular representations projecting conventional gender identities, and also by positioning himself counter to the material gender pattern of location around him. From his viewpoint, in this Japanese corporation for which he worked, 'ninety-nine-point-nine per cent' of men worked on the engineering side and no men, so far as he was aware, worked as home economists. On the other hand, in a couple of cases elsewhere in the world-wide organization, women filled this microwave product planning role. Even outside, in the world of ordinary people who buy and use microwave ovens, he saw men being the ones concerned with 'the power, the construction, the bells and whistles' on the oven while women were the ones concerned with what in fact concerned him also: its practical utility in the kitchen.

These two instances drawn from Electro UK show a woman and a man whose sense of self involves an appropriation of technology that *negotiates with* rather than adopts the dominant projected gender identities and the dispositions suggested by the gender pattern of location in the technological sphere. They exemplify the relationship between structure and agency, the given and the chosen. Frigga Haug calls this 'the intertwining of processes of self-fulfilment with the fulfilment of cultural expectations'. She also observes that

> human beings do not simply fulfil norms, nor conform in some uncomplicated way; that identities are not formed through imitation, nor through any simple reproduction of predetermined patterns, but that the human capacity for action also leads individuals to attempt to live their own meanings and find self-fulfilment, albeit within a predetermined social space. (Haug, 1987: 42)

Karen and Keith were part-consciously making their choices in response to a feminist gender symbolism. The prevailing gender symbolism and gender structure, however, even if not fully determining, are persuasive enough to carry the majority along, and these two people were both aware of a difference from most others of their sex.

The Commercial Uses of Gender Innovation

We saw that Home-Tec and Wonderworld, the retail 'multiples' selling electrical consumer goods of the brown and white kind, were characterized by somewhat more innovatory gender locations and representations than Electro, the site of the microwave's manufacture. They had Equal Opportunities policies in employment, they had promoted more women to management (albeit few to *store* management) and were trying through training strategies to break down the stubborn tendency to sex-typing that labelled the selling of brown goods and white goods as respectively masculine and feminine.

Several possible prompts to a such a counter-sexist strategy spring to mind, to all of which the retailer is more exposed than the manufacturer – although they also reach the manufacturer in diluted form. First, women themselves, in the wider society, are making their own movement and pressing their own expectations: of equal opportunities, equal pay, equal treatment. The few women in Electro, the rather more (though still not many) women in Home-Tec and Wonderworld, who are doing responsible and reasonably paid jobs are there partly because they aimed to be there and overcame any impediments in their way. Women like this, besides, are those whom the manufacturer and retailer of electrical consumer durables must please with their goods and their sales promotions. The retailer is closer than the manufacturer, and thus more sensitive, to changes in the customer base of its stores, which could be one reason for their greater openness to change in the gender pattern of location.

Secondly, cost advantages of cheap female labour notwithstanding, a unisex workforce is more flexible to deploy than a sex-segregated one. At present no qualified women engineers or technical managers present themselves to Electro UK; but increasingly young men do turn up prepared to take assembly line jobs and just a few young women employed as operators surprise management by appearing to aspire to technical skills. Electro can be seen as responding pragmatically to such circumstances. Retail business by contrast consists predominantly of many kinds of selling, buying, marketing and related jobs, for all of which more and more women, as time goes by, are as well or better qualified than men. Home-Tec and Wonderworld for their part have been learning this lesson.

We mentioned in Chapter 4 a video that Wonderworld had made to help encourage women to develop brown goods product knowledge as a step towards more flexible deployment of labour on the sales floor. The video shows very clearly how women's own interests and those of the company coincide at this point. It also demonstrates representational work simultaneously reshaping feminine gender identity while reproducing the gender relation as asymmetrical.

The young sales assistant chosen to play the lead role in the video is Toni. She has a working-class accent, is trim, pretty, bright and assertive. A joky voice-over introduces her with the following words:

> She's been working for Wonderworld for about twelve months. She enjoys selling and immediately formed attachments to lots of the products: particularly vacuum cleaners. Like lots of ladies, or for that matter fellows, who work at Wonderworld, Toni likes to sell products she knows most about and feels confident with. Washing machines and dishwashers came naturally. And refrigeration was, in Toni's words, 'easy-peasy'. . .

> [One day] Toni approached the sales floor and she noticed it was particularly busy on the TV and hi-fi side. Lots of customers were looking at CDs and two or three different groups were clearly needing help with information on video recorders. . .
>
> There was only one other salesperson on that side of the store that day, and as he was on his own, Toni took the plunge and entered the foreboding world of high technology.

Needless to say, Toni comes seriously unstuck with a demanding customer and appeals to the training officer for a course in product knowledge – which is the prompt for the ensuing footage on the features and benefits of VCRs.

This training video projects two messages about gender simultaneously. Toni is being remade as the *technologically competent woman*. Yet the authoritative voice-over in the film is male. The training officer and manager whose superior know-how is mobilized to solve Toni's problem are both male, and certain devices, no doubt unconscious, confirm both their masculinity and authority: the men are much taller, hold a clipboard, address the camera direct. Toni is framed by the men, literally 'looks up to them'. There are, besides, two touches of innuendo in the video, that tend to sexualize Toni. They arise within the actual social relations of the performers – who are of course not professional actors but Wonderworld personnel who clearly know each other and will be personally known to the video's anticipated audience. At one point reference is made to a potentially romantic involvement with a salesman. At another, the training manager says, 'You sound like my kind of girl.' Toni, to get the dealings back on line says, '*No, but seriously. . .*' and continues to explain her problem.

At one level, then, femininity is being reshaped in this video to allow for the integration of knowledge formerly considered masculine. Toni wants this. So do her employers. Toni's slightly androgynous name may be that of the real saleswoman who acts the part or it may have been chosen by the video-producers. Either way, it eases this metamorphosis. At another level, however, a message is conveyed that Toni remains, for all this, very much a woman, available to men, subject to their relative authority and dependent upon their greater technical knowledge.

Identities on Offer

If pressure from women themselves and the employers' interest in the most profitable exploitation of labour power are two forces for gender innovation, a third is to be found in capitalist market processes. Sales of consumer goods are increased by any tendency to individualize the consumer. A woman who sees herself as having her own life, her own money, an autonomous subjectivity, may be expected to spend more on consumer goods than one who is 'only' a daughter or a wife – while at the same time spending no less on those functions of the family for which she still may (in fact) remain responsible. The retailer of consumer goods sees benefit today in addressing, constructing and, so far as possible, controlling the behaviour of men and women as *differentiated individuals*, likely to set up their own households, open their own bank accounts and experience autonomous needs. The advertising and retailing trend towards identifying and appealing to 'lifestyle' represents just this turn towards a postmodern multiplicity of endlessly differentiated wants waiting to be satisfied. Conventional sex-stereotyping, by contrast, limits the potential market to two types – worse, a couple buying as a unity. 'Capital has fallen in love with difference,' said Jonathan Rutherford (1990a: 11).

Advertising of microwave ovens, microwavable foods and electrical consumer goods more generally, demonstrates the way capital experiments with projections of gender identity in pursuing its interests – enhanced sales. We draw here on analysis of recent advertisements and commercials and three interviews, two with advertising executives responsible for the promotion of microwaves and other electrical consumer products, and one with a marketing executive for a new microwavable food product.

Advertisers are never insensitive to gender. At times they appear to evade gender representations. For example, some advertisements for microwave ovens show the oven alone in a context devoid of human beings, and promote its 'features, advantage and benefits' (FAB). When this tactic is chosen it is precisely to avoid what at times seems a gender minefield. Advertisers are only too aware that, while they construct and project gender identities, people are moving on and the advertiser can all too easily hit a false note, alienate the potential customer and damage sales. Besides, any advertising representation of a woman or man is inevitably culturally specific, a 'type', which may limit sales to a narrower target group than is at best attainable by the product. It can sometimes be the safest course, therefore, to show nothing but the technologically clever machine itself, its multidirectionality and socially interpretive flexibility unrestrained.

On the other hand, this strategy is not adequate if what the manufacturer suspects is that the target group is failing to identify itself as customer, if what is needed is the persuasive social confection of a user. Microwave manufacturers have tentatively approached this user in several different interpretive sketches. Where women have been used, they have tended to be represented as cooks in the domestic, particularly family, context. When men have been used they have been in one of three guises: the appreciative eater of microwave foods, the interested recipient of microwave technology, or the expert professional chef. An example of the first is an advert for a microwave with browning facility which showed two men holding sausages on forks in front of their faces, one sausage pale and flabby, downward bending, the other crisp, brown and upturned in a smile. The text begins: 'When you microwave a sausage, you should get a banger, not a damp squib.' There was a humorous phallic sub-text here. The second male guise is illustrated by a father and son standing beside their new microwave, studying the instruction manual. The third approach, masculine professionalism, was used to promote a very advanced model. French chefs were employed to suggest that this was the most sophisticated combination oven on the market, fully capable of *haute cuisine*.

What has remained constant in the representation of women in connection with microwaving is the understanding that it is they who maintain the responsibility for choice of diet, for cooking well, for quality of provision for all the family. The woman is the one who knows best. This persona was deftly drawn in one particular television commercial for a microwavable line of foods. First a child, then a man, place a box containing an unnamed ready-meal in the microwave. Each time, to the bemusement of the family, the microwave spits it out, flings it across the room. Then the woman, with a knowing smile to camera, places the correctly branded ready-meal in the oven, which receives it appreciatively. 'Yum, yum,' it flashes on its display panel. Mother knows best. The thinking behind such imagery, a marketing manager for such products explained, is that men and children are seen as opening the family to change, the tip of a wedge to open up to new ways the traditional cooking and eating practices of which the housewife and mother is guardian:

> Men are great experimentalists, seen as that, in cooking. They're more likely for example, at the shelves, on a shopping trip, to pick up things.

But it is the woman to whom the main appeal must be made.

> Unless the housewife says it's safe to the family, the family won't believe it, that's the point. So you've got to convince her first. She probably convinces her family and then is prepared to let her family take over and use, as and when. The bridging of that gap is still very much in her hands. So we're still educating *her*.

While advertisers in their commercials project families that suit their needs, family and household structures really are, as we have seen, undergoing change. Women and men, actors in the microwave-world, are finding themselves in new circumstances and subjectivities are shaping up differently. People are going on holiday to Thailand and Mexico, coming back and spreading a taste for Thai coconut curry and *enchiladas*. Teenagers are getting more demanding and want the meals they fancy, just when they want them. Women are struggling to run both a job and a household without cracking up or entirely submerging their own identities. More women and men are living alone, facing the problem of sustaining an interesting and varied diet in single portions. If their clients are not to look foolishly out of touch, advertisers must respond to this reality. In responding, they also, of course, put on offer to the viewer a persuasive range of proposed identities and so help to construct the world they purport to describe. There is an in-built conveyor-belt of change in consumer markets and marketing. As one advertiser put it, 'there's probably a three- or four-year life for the man or the woman you see most in advertising. Then their very exposure kills them off and you have to move on.'

'Falling in love with difference', then, is something both individuals have done and capital has fantasized. And here a clear distinction must be made between differences within a gender and difference between the (two) genders. Advertisers have espoused the former, have modified their adherence to gender stereotypes in what they call a 'lifestyle' approach. One advertiser explained how *within* male voices and female voices there are distinct differentiating tones. Among the latter for instance there are sexualized women for some situations, emphatic women, housewifey women, mumsy types for others. Yet there is no sex-neutrality in advertising. The emphasis on difference *between* genders, gender as relational, is uninterrupted.

The autonomous, income-earning woman is a case in point. In the advertising of some products, perfume for instance, she is free of family responsibilities. But she is distinctively woman, and perfumes are clearly gendered, his and hers. The new woman in the family is less likely today to be the boring slave to her husband's shirts. She is still half of the domestic couple, still identified with the domestic sphere, but now she is clever, humorous, and wryly rebellious. She is portrayed as before as the one who carries responsibility for quality of life. Today, however, she copes with the many more demands arising from the increased individuality of family members. She is rewarded by the advertisers with a modernized identity. Now (as one put it) she 'is the person who always has the last laugh. . . And that's a real move on from six or seven years ago when the mum wanted nothing more than to prepare the perfect meal.'

The advertisers' recent masculine creations likewise are unmistakably men, though their relationship to women and the family is subtly differentiated and adjusted. On the one hand some versions of manhood have quit the family. For example, 'there was the fellow that obviously lives down Canary Wharf, Docklands,' an advertiser recalled, 'who wandered down on a Sunday morning and got his paper, had an independent life, wasn't about family. . .' On the other an updated model of family man has made an appearance. The sophisticated public is believed now to perceive the stereotypical dad with son and car as boring. That was all about 'buying this thing so that you can be seen within your family to be the person who's delivered success to it as the head of the family'. Today's men are preferred

> to take pride in different and good relationships. Spending time with your family, having a good relationship with your son, is an individual thing now. It's the thing about quality time, isn't it? It's a very personal thing. It's not aspirational any more.

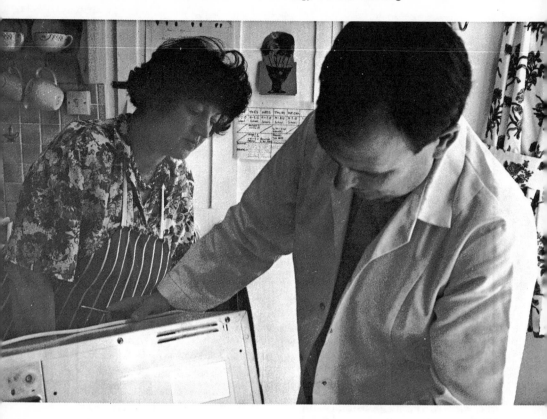

Despite the flowering of women and men into different lifestyle images and modernized heterosexual relations, the continuity of gender as such, its relational essence, continues to be emphasized and exploited. If a woman is used, the implication is that only women are addressed. Thus one advertiser told us: 'it would be a bit incongruous if you had a woman in an ad with camcorders *alone*'. A man may sometimes stand for 'people', when he is depicted alone. But in a situation where both sexes are present their asymmetry must be reiterated.

> It would be folly to present a commercial probably where if the woman was present, she was *not* the one doing the work in the kitchen. . . it just wouldn't be credible, because, you know, it's an area of excellence which they are meant to understand.

Where a woman or a man are used in a cross-gender context it is always with a gender purpose. For example, where a woman is shown using a camcorder it may be to demonstrate that it is especially small and light. Where a man is shown using a microwave, either it is because he is a professional chef and the advert emphasizes technological complexity, or it is because he is 'a not-woman' and the advert bears the meaning 'any fool can do it'.

Technology, Gender and the Feminist Project

Doing, making and producing and the tools that aid those activities, are, we have argued, what is meant by technology in the full sense of the word. That has justified our inclusion of cooking within the scope of technology in this book. In common usage however the word 'technology' conveys a more limited concern: artifacts, processes and knowledge used in production (and organized destruction) involving *engineering*. Thus technology is split in two. There is a relatively clearly defined sphere, Technology-with-a-capital-T, that has developed as masculine. And there is a diffuse residual sphere of doing and making whose activities are variously gendered (amateur fishing as masculine, for instance, typing as feminine). The most characteristic *domestic* doings are feminine. In the home, utensils, tools and machines are used but not as a rule made. Most (like the microwave oven) are *engineered* in the sphere of Technology and sold as commodities. Technology-with-a-capital-T however involves both making *and* using equipment, tools and machines.

The association of Technology symbolically with masculinity and materially with men, in the location of the male sex close to Technological work, skills and machines, has fostered a high relative value in all three. These are not the only linked terms in this chain of meaning of course. Men and masculinity are also associated with other highly valued attributes such as courage and reason. Technology is valued not only because it is masculine but because of its often powerful and visible effects (supersonic flight, mass production). Nonetheless the mapping of men/masculinity on to Technology and women/femininity on to non-Technology has been an important aspect of the iterative processes that have produced both men and Technology as *relatively important*.

When, during the three years of this research, we told people we were engaged on a sociology of the microwave oven, it invariably drew a smile. At first this was unnerving. We began to smile ourselves, disarmingly, in anticipation. It made us feel a little shamefaced, apologetic: this could not be serious sociology. Then we remembered that nobody had smiled when the subject of research had been computer-aided design or nuclear magnetic resonance scanning. And the penny dropped. The smiles were precisely a part of our research material. They said, in effect, microwave = domestic = feminine = unimportant.

The material and representational effects of the technology/gender relation, as we have seen it expressed in this unimportant microwave-world, furnish the circumstances and images that dispose and propose the likely self-identities of its actors, male and female. Its men are more likely to identify themselves as having *agency*, and to be valued by others for that. Its women are more likely to find their sense of self in, and be seen as, subsidiary actors concerned with *sustenance*. Men's doing and making seems more important, effective and far-reaching; women's

seems less important, repetitive, a matter of provisioning and maintaining. Men's lives seem a project, women's a cycle.

It is not that women and the supportive world of the domestic are considered altogether *un*-important. Indeed their worth is recognized by men and women alike. Men love women *in their place*, and few would deny their indebtedness to women for their daily dinners, microwaved or otherwise. Women and the domestic, however, are valued strictly within their proper location: women separate from men in supportive roles and in the domestic sphere; the domestic sphere clearly distinct from and subordinate to the public world. No cooking smells in the office, please. We are speaking, then, only of *relative* value. It is acknowledged that agency and sustenance, these two facets of identity, indeed of life, thrown into relief by the technology/gender relation, are both important. But our culture has cast them in a reverse relation to the one they should ideally have. Agency should serve sustenance, engineering should be directed towards a sustainable everyday life, production should facilitate reproduction. Instead, family life is drained by the workplace, everyday life is exploited and deformed by Technology, and the gift of human agency risks becoming an arrogant masculine project of transcendence that owes no responsibility to care.

Equally damaging, this pattern of relations typecasts male and female subjectivities in the straitjacket of a dichotomy in which most women lack agency while most men fail to nurture. As in all such gender binarisms this splitting tends to complementarity and asymmetry. One gender becomes the missing half of the other, men and women stand in for each other in half their potential life-space. Complementarity assures everyone's impoverishment. Asymmetry writes down the feminine: it is held to be more serious a problem that women lack agency than that men lack an orientation to sustenance. Official sex equality policies acknowledge the former, only feminism argues the latter.

Coming back from an excursion into the microwave-world we can perhaps see more clearly the complexity of feminist projects for change in the technology/gender relation.

Partly, women's perceived distance from Technology is a question of blindness. Many women do have a valuable understanding of engineered technologies by virtue of having been at the receiving end of them – the ones to operate the machines, use the commodities. The knowledge a woman has gained about a microwave oven from using it has *worth*, as the engineer's knowledge of it has worth. First, then, women's existing knowledge about Technology should be recognized.

We do, however, also need to ensure that women have less costly access to recognized technological skills and qualifications. Several initiatives have been tried. Women have set up access courses in which woman students, taught by women teachers, can get a grip on engineering, computer systems analysis and other subjects dominated by men without having to outface a masculine culture before they have the confidence to do so. A few institutions offering mainstream engineering courses have tried to adapt their curricula to make them more meaningful to women, and have provided social support for the small minority of women among students and staff. Such initiatives have barely scratched the surface of the problem, however, and much more thought, effort and above all financial backing is needed if women are to feel they can grasp Technology without damaging their own lives. An ever greater proportion of education and research spending is on science and technology. Women are missing out on grants, jobs and rewarding careers.

More importantly, only if women have the key skills and the key jobs can we interrogate technological developments, assess them against various women's needs and influence their course. Sandra Harding has called women 'valuable strangers' to the social order, capable of seeing it with new eyes, questioning all that has been taken for granted (Harding, 1991: 124). Women have this power of the 'valuable stranger' to Technology, and from the standpoint of our own struggles we can see what needs to change. Women need technologies that are *for* women, not technologies imposed on them by a technoscience system that has no knowledge of the lives of different women in different countries, and cares little about them. A feminist standpoint arises through women's conscious struggles, and in this sense women's active opposition in many countries and continents to nuclear weapons, genetic engineering, environmental destruction and many other products of technoscience are a defining feature of contemporary feminisms and are shaping new identities for women.

Apart from professional levels of skill, there is another kind of know-how, less specialized, that women need. Not all women would want to be professional engineers – why should they? But no women should feel themselves at a disadvantage, as they often do today, for lack of a basic understanding of engineered technologies. So closely are our bodies and our identities tied up with technology in today's world, we are all, as Donna Haraway has reminded us, 'cyborgs' (Haraway, 1991). Without our clothes and equipment, our medicines and appliances, our systems of shelter, heating and transport, we would not be the people we think of ourselves as being. If women are to control their own lives, they must have an everyday knowledge of how things work. Girls could acquire this effortlessly at a young age, were it not that masculinist culture of home and school stand in their way.

What this study has shown is the tendency for Technology to be accorded a status out of all proportion to its actual importance. Men as a sex are at risk, because of the way their masculinity is shaped in proximity to technology, of giving unquestioning priority to the technological project. Women have one advantage: their feminine gender identity positions them outside the magic radius of Technology's appeal. We have to find the tools for our struggle within the gender relations we have inherited, and the keenest liberatory tools femininity affords us are scepticism of the hype surrounding Technology, and respect for other kinds of doing and making. We need to continue to recover and restate, as the women's movement always has, the value of the things women traditionally do, technologies with a little t. The feminine is not a terrain from which we should try to escape altogether, but one we should keep a foot in, landscaping it anew to suit our varied and changing needs as women. What is more, it is terrain we have to draw men on to. Women cannot simply say of the domestic, the routine, the familial, the everyday, 'This isn't *all* I am.' We have to make sure men hear us saying as well, 'Hey! this isn't only *mine*'.

References

Albury, David and Schwartz, Jo (1982) *Partial Progress: The Politics of Science and Technology*. London: Pluto Press.

Allen, M. (ed.) (1990) *The Times 1000, 1990–91*. London: Times Books Ltd.

Andrews, Gordon (1990) 'Developments in microwave oven design', unpublished paper.

Angela, Frances (1990) 'Confinement', in Rutherford (1990b).

Arnold, Erik and Burr, Lesley (1985) 'Housework and the appliance of science', in Faulkner and Arnold.

Association of Manufacturers of Domestic Electrical Appliances, London (1991), statistics supplied on request.

Barrett, Michèle (1980) *Women's Oppression*. London: Verso.

Bereano, Philip, Bose, Christine and Arnold, Erik (1985) 'Kitchen technology and the liberation of women from housework', in Faulkner and Arnold.

Berg, Anne-Jorunn (1990) 'He, she and I.T.: designing the home of the future', in *Technology and Everyday Life: Trajectories and Transformations*, Report No.5, 28–9 May, Norwegian Research Council for Science and the Humanities.

Bertell, Rosalie (1985) *No Immediate Danger: Prognosis for a Radioactive Earth*. London: The Women's Press.

Bijker, Wiebe E., Hughes, Thomas P. and Pinch, Trevor J. (eds) (1990) *The Social Construction of Technological Systems: New Directions in the Sociology and History of Technology*. Cambridge, Mass: MIT Press.

Bloor, D. (1976) *Knowledge and Social Imagery*. London: Routledge & Kegan Paul.

Bose, Christine E., Bereano, Philip L. and Malloy, Mary (1984) 'Household technology and the social construction of housework', *Technology and Culture* 25 (1), January: 53–82.

Bradley, Harriet (1989) *Men's Work, Women's Work: A Sociological History of the Sexual Division of Labour in Employment*. Cambridge: Polity Press.

British Market Research Bureau (1990) *Target Group Index*. London: British Market Research Bureau.

Burnett, Sally-Ann (1990) 'The social, technical and nutritional implications of increased use of microwave energy in food preparation', PhD thesis, University of Wales, Cardiff.

Butler, Judith (1990) *Gender Trouble: Feminism and the Subversion of Identity*. London: Routledge.

Callon, Michel (1986) 'The sociology of an actor-network: the case of the electric vehicle', in Michel Callon et al.

Callon, Michel and Latour, Bruno (1981) 'Unscrewing the big Leviathan: how actors macro-structure reality and how sociologists help them do so', in Knorr-Cetina and Cicourel.

Callon, Michel and Law, John (1982) 'On interests and their transformation: enrolment and counter-enrolment', *Social Studies of Science*, 12: 615–25.

Callon, Michel, Law, John and Rip, Arie (eds) (1986) *Mapping the Dynamics of Science and Technology*. Basingstoke: Macmillan Education.

Carter, Ruth and Kirkup, Gill (1990) *Women in Engineering: A Good Place to Be?* Basingstoke: Macmillan Education.

Chabaud-Rychter, D., Fougeyrollas-Schwebel, D. and Sonthonnax, F. (1985) *Espace et temps du travail domestique*. Paris: Meridien-Klincksieck.

Charles, Nickie and Kerr, Marion (1988) *Women, Food and Families*. Manchester: Manchester University Press.

Chisholm, Lynne and Holland, Janet (1986) 'Girls and occupational choice: anti-sexism in

180   *Gender and technology in the making*

action in a curriculum development project', *British Journal of Sociology of Education*, 7 (4): 353–66.

Cockburn, Cynthia (1983) *Brothers: Male Dominance and Technological Change*. London: Pluto Press.

Cockburn, Cynthia (1985) *Machinery of Dominance: Women, Men and Technical Know-how*. London: Pluto Press.

Cockburn, Cynthia (1987) *Two-Track Training: Sex Inequalities in the Youth Training Scheme*. Basingstoke: Macmillan Education.

Collins, H.M. (1981) 'Stages in the empirical programme of relativism', *Social Studies of Science*, 11: 3–10.

Cowan, Ruth Schwartz (1985) 'How the refrigerator got its hum', in MacKenzie and Wajcman.

Cowan, Ruth Schwartz (1989) *More Work for Mother: The Ironies of Household Technology from the Open Hearth to the Microwave*. London: Free Association Books.

Deem, Rosemary (1986) *Schooling for Women's Work*. London: Routledge & Kegan Paul.

Delphy, Christine (1979) 'Sharing the same table', in Harris, Chris.

Department of Education and Science (1987) *The National Curriculum 5–16: a Consultation Document*, London: DES (July).

Department of Education and Science (1989) *National Curriculum: From Policy to Practice*, London: DES.

Economist Intelligence Unit (1990) 'Quarterly product review: furniture, electrical appliances and sound equipment', *EIU Retail Business*, 386: 3–29.

EIAJ (Electronic Industries Association Japan) (1990) *Investment in Britain by Japan's Electronic Industry*. London: Burrup Mathieson.

Epstein, Cynthia Fuchs (1988) *Deceptive Distinctions: Sex, Gender and the Social Order*. New Haven: Yale University Press; London: Russell Sage Foundation.

Faulkner, Wendy and Arnold, Erik (eds) (1985) *Smothered by Invention: Technology in Women's Lives*. London: Pluto Press.

Fraser, Nancy and Nicholson, Linda J. (1990) 'Social criticism without philosophy: an encounter between feminism and postmodernism', in Nicholson, Linda J.

Game, Ann and Pringle, Rosemary (1983) *Gender at Work*. North Sydney: George Allen & Unwin.

Gershuny, Jonathan and Robinson, John P. (1988) 'Historical changes in the household division of labor', *Demography*, 25 (4) November. pp. 537–52.

Glucksman, Miriam (1990) *Women Assemble: Women Workers and the New Industries in Inter-war Britain*, London: Routledge.

Grint, Keith and Woolgar, Steve (1992) 'Computers, guns and roses: what's social about being shot?', *Science, Technology and Human Values*, 17 (3): 366–80.

Hacker, Sally (1989) *Pleasure, Power and Technology: Some Tales of Gender, Engineering and the Cooperative Workplace*. Boston, Mass.: Unwin Hyman.

Haraway, Donna J. (1991) *Simians, Cyborgs and Women: The Reinvention of Nature*. London: Free Association Books.

Harding, Sandra (1986) *The Science Question in Feminism*. Milton Keynes: Open University Press.

Harding, Sandra (1991) *Whose Science? Whose Knowledge?: Thinking from Women's Lives*. Milton Keynes: Open University Press.

Harris, Chris (ed.) (1979) *The Sociology of the Family, Sociological Review* Monograph No. 28.

Haug, Frigga (ed.) (1987) *Female Sexualization*. London: Verso.

Higgins, Tracey and Almond, Abi (1989) *Project MOZ: Microwave Oven Qualitative Research Final Report*. London: Association of Manufacturers of Domestic Electrical Appliances.

Holland, Janet (1986) 'Gender and class: adolescent conceptions of the social and sexual division of labour', *CORE*, 10 (1).

Holland, Janet (1987) *In Search of Meanings: Girls and Occupational Choice*, Working Paper No. 10, Institute of Education, London.

hooks, bell (1982) *Ain't I a Woman? Black Women and Feminism*. London: Pluto Press.

Johansson, Birgitta (1988) *Ny Teknik och Gamla Vanor*. Linköping, Sweden: Linköping University Press.

Kelly, A. (1981) *The Missing Half: Girls and Science Education*. Manchester: Manchester University Press.

Kessler, Suzanne J. and McKenna, Wendy (1978) *Gender: an Ethnomethodological Approach*. New York: John Wiley and Sons.

Keynote Report (1990) *An Industry Sector Overview: Household Appliances (White Goods)*, 7th Edition. Middlesex: Keynote Publications.

Kling, Rob (1991) 'Computerization and social transformations', *Science, Technology and Human Values*, 16 (3): 342–67.

Kling, Rob (1992) 'Audiences, narratives and human values in social studies of technology', *Science, Technology and Human Values*, 17 (3): 349–65.

Knorr-Cetina, K. and Cicourel, A.V. (eds) (1981) *Advances in Sociological Theory and Methodology*. London: Routledge & Kegan Paul.

Latour, Bruno (1986) 'Powers of association', in Law, J.

Latour, Bruno (1987) *Science in Action*. Milton Keynes: Open University Press.

Latour, Bruno and Woolgar, Steve (1979) *Laboratory Life: The Construction of Scientific Facts*. London and Beverley Hills: Sage Publications.

Law, John (ed.) (1986) *Power, Action and Belief: A New Sociology of Knowledge*. London: Routledge & Kegan Paul.

Law, John (1989) 'Technology and heterogenous engineering: the case of Portuguese expansion' in Bijker et al.

Lazell, David (1981–2) 'Home economics for male students', *Home Economics*, December 1981/January 1982. pp. 21–2.

Lazonick, William H. (1979) 'Industrial relations and technical change: the case of the self-acting mule', *Cambridge Journal of Economics*, 3: 231–62.

Leacock, Eleanor Burke (1981) *Myths of Male Dominance*. New York and London: Monthly Review Press.

Lloyd, Genevieve (1984) *The Man of Reason: 'Male' and 'Female' in Western Philosophy*. London: Methuen.

Lorber, Judith and Farrell, Susan A. (eds) (1991) *The Social Construction of Gender*. Newbury Park and London: Sage Publications.

MacKenzie, Donald (1981) 'Interests, positivism and history', *Social Studies of Science*, 11: 498–504.

MacKenzie, Donald and Wajcman, Judy (eds) (1985) *The Social Shaping of Technology*. Milton Keynes: Open University Press.

McNeil, Maureen (ed.) (1987) *Gender and Expertise*. London: Free Association Books.

Marketpower Ltd (1989) *Microwave Foods Europe*. London.

Marketpower Ltd (1991) *Microwave Oven Data*. London.

Martin, Michele (1991) *'Hello, Central?' Gender, Technology and Culture in the Formation of Telephone Systems*. Montreal and Kingston: McGill-Queens University Press.

Mead, Margaret (1935) *Sex and Temperament in Three Primitive Societies*. London: Routledge.

Morgall, Janine Marie (1991) *Developing Technology Assessment: A Critical Feminist Approach*. Lund, Sweden: University of Lund.

Murcott, Anne (1983a) 'Women's place: cookbooks' images of technique and technology in the British kitchen', *Women's Studies International Forum*, 6 (1): 33–9.

Murcott, Anne (ed.) (1983b) *The Sociology of Food and Eating*. Aldershot: Gower.

Newton, Peggy (1987) 'Who becomes an engineer? Social psychological antecedents of a non-traditional career choice' in Spencer and Podmore.

Nicholson, Linda J. (ed.) (1990) *Feminism/Postmodernism*. New York and London: Routledge.

Noble, David (1977) *America by Design: Science, Technology and the Rise of Corporate Capitalism*. Oxford: Oxford University Press.

Noble, David (1984) *The Forces of Production: A Social History of Industrial Automation*. New York: Alfred A. Knopf.

Nyberg, Anita (1989) *Tekniken – Kvinnornas Befriare?*, Linköping, Sweden: Linköping University Press.

Oakley, Ann (1972) *Sex, Gender and Society*. London: Temple Smith.

Oakley, Ann (1974) *The Sociology of Housework*. Oxford: Martin Robertson.

Ortner, S. and Whitehead, H. (eds) (1981) *Sexual Meanings: The Cultural Construction of Gender and Sexuality*. Cambridge: Cambridge University Press.

Pateman, Carole (1988) *The Sexual Contract*. Cambridge: Polity Press.

Phillips, Anne and Taylor, Barbara (1980) 'Sex and skill: notes towards a feminist economics', *Feminist Review*, 6: pp. 79–88.

Pinch, Trevor J. and Bijker, Wiebe E. (1990) 'The social construction of facts and artifacts: or how the sociology of science and the sociology of technology might benefit each other', in Bijker et al.

Rosaldo, M.Z. and Lamphere, L. (eds) (1974) *Women, Culture and Society*. Stanford: Stanford University Press.

Rose, H., McLoughlin, I., King, R. and Clark, J. (1986) 'Opening the black box: the relation between technology and work', *New Technology, Work and Employment*, 1 (1): 18–26.

Rutherford, Jonathan (1990a) 'A place called home: identity and the cultural politics of difference', in Rutherford (1990b).

Rutherford, Jonathan (ed.) (1990b) *Identity: Community, Culture, Difference*. London: Lawrence & Wishart.

Sale, Roger (1988) 'Women and industrial design – breaking the mould', *New Home Economics*, December 1987–January 1988. pp. 18–19.

Shiva, Vandana (1989) *Staying Alive: Women, Ecology and Development*. London: Zed Books.

Spencer, Anne and Podmore, David (eds) (1987) *In a Man's World: Essays on Women in Male-Dominated Professions*. London and New York: Tavistock Publications.

Stanworth, Michelle (ed.) (1987) *Reproductive Technologies: Gender, Motherhood and Medicine*. Cambridge: Polity Press.

Thrall, Charles A. (1982) 'The conservative use of modern household technology', *Technology and Culture*, 23 (2), April: pp. 175–94.

Wajcman, Judy (1991) *Feminism Confronts Technology*. University Park, Pennsylvania: Pennsylvania State University Press.

Walby, Sylvia (ed.) (1988) *Gender Segregation at Work*. Milton Keynes: Open University Press.

Walden, Louise (1990) *Genom Symaskinens Nalsoga: Teknik och Social Forandring i Kvinnokultur och Manskultur*. Stockholm: Carlssons.

Weedon, Chris (1987) *Feminist Practice and Poststructuralist Theory*. Oxford: Basil Blackwell.

Weeks, Jeffrey (1990) 'The value of difference', in Rutherford (1990b).

Weinrich-Haste, Helen and Newton, Peggy (1983) 'A profile of the intending woman engineer', *Equal Opportunities Commission Research Bulletin*, 7.

Wilson, Edward O. (1975) *Sociobiology: The New Synthesis*. Cambridge, Mass.: Belknap/Harvard University Press.

Winner, Langdon (1980) 'Do artifacts have politics?', *Daedalus*, 109: 121–36.

Winner, Langdon (forthcoming) 'Upon opening the black box and finding it empty: social constructivism and the philosophy of technology', *Science, Technology and Human Values*.

Woolgar, Steve (1981) 'Interests and explanation in the social study of science', *Social Studies of Science*, 11: 365–94.

Index